ECONOMIC, TECHNOLOGICAL AND LOCATIONAL TRENDS IN EUROPEAN SERVICES

To Denise

Based on a report prepared for the Commission of the European Communities, Directorate-General for Science, Research and Development, as part of the FAST Programme (Forecasting and Assessment in the Field of Science and Technology).

Economic, Technological and Locational Trends in European Services

JEREMY HOWELLS

Research Fellow
Centre for Urban and Regional Development Studies
The University of Newcastle upon Tyne

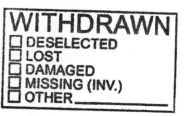

Avebury

Aldershot · Brookfield USA · Hong Kong · Singapore · Sydney

Publication No. EUR 11148/I of the
Commission of the European Communities,
Directorate-General Telecommunications, Information
Industries and Innovation, Luxembourg

Published by

Avebury

Gower Publishing Company Limited
Gower House
Croft Road
Aldershot
Hants GU11 3HR
England

Gower Publishing Company
Old Post Road
Brookfield
Vermont 05036
USA

ISBN 0 566 056461

Printed and bound in Great Britain by
Biddles Limited, Guildford and King's Lynn

Contents

vi

Acknowledgements

This book draws upon a research project funded by the Forecasting and Assessment Council in the Field of Science and Technology (FAST) II, Directorate-General for Science, Research and Development, Commission of the European Communities (Contract No. FST-061-UK), and the UK Economic and Social Research Council.

The author wishes to acknowledge all those people whose help and advice has been drawn upon in writing this report : from the European Commission Oliver Ruyssen, from DG XII, and Sven Illeris, during his times as FAST Fellow, and in particular Ian Clark, from DG XVI, who co-ordinated the mapping of service employment and provided additional data. Within CURDS itself I would like to thank Mark Hepworth, who gave particular encouragement and help, David Charles, John Sanderson and Andrew Gillespie who also made valuable contributions; whilst John Goddard with Alfred Thwaites provided overall advice and co-ordination on the project. A special thanks goes to Sue Robson and Denise Rainford-Weites for their typing and secretarial assistance associated with the project.

1 Trends and prospects for European services

Introduction and Background

This book on services and technological development in Europe arises out of a research programme that the Centre for Urban and Regional Development Studies (CURDS) has undertaken for the Forecasting and Assessment in the field of Science and Technology (FAST) II initiative funded by the European Commission, under the direction of the Directorate – General for Science, Research and Development (DG XII) and the United Kingdom (UK) Economic and Social Research Council. This study, as such, is based on a research report submitted to the Commission (Howells 1987a) which in turn is closely associated with an earlier UK report (Howells and Green 1986a) in the sense that it broadens the issues of the UK study and incorporates other European research contributions and material. An updated and extended version of the UK report has also been published by Gower (Howells and Green 1988).

The other European research contributions noted above, arise from a number of sources. These include two seminars CURDS organised in conjunction with the Commission. The first in May 1985 between the services and regional development research sub-group (SERV6) and other FAST II research teams and the second in October 1985 which included researchers on the FAST II SERV6 Research Network. Another source was from five other research organisations who prepared reports under the SERV6 sub- programme (Netherlands Economic Institute, 1986; Guerin and Outrequin, 1986; CENSIS 1985; Cuadrado 1986; and Pederson 1985; together with short reports by Bailly 1985; Cappellin 1985; Lazzeri et.al. 1985; Lambooy and Renooy 1985)

In addition a questionnaire was sent out to national experts selected from the FAST II-SERV6 Network which would provide their evidence and views relating to services, technological innovation and regional development. Some of this of evidence has been incorporated into this book (however see Howells 1987a, 94-109 for a more formal review of this material). Finally, CURDS has continued to keep in contact with FAST II sub-programmes by attending meetings and exchanging material. All these different initiatives have been in close consultation with Sven Illeris, the FAST Fellow co-ordinating the programme.

However, unfortunately a large amount of research evidence which is being generated in other FAST II sub-programmes is only now being available, whilst it has been possible to only briefly cover some of the research issues raised in detail by the other SERV6 documents. Readers are therefore directed to study the wealth of information contained in these reports (available as FAST Occasional Papers) which are meant to be 'stand-alone' pieces of research.

A key element in the overall programme is the identification and formulation of broad policy issues and strategies for the Community in relation to services. As such Chapter 6 of the book focuses on the long term policy issues arising from the two studies. In this respect CURDS has been responsible for providing advice to the FAST II team within

DGXII, and other Commission agencies, in the formulation of Community strategy in this field.

Scope of the Book

The rest of the book is structured the following way. Chapter 2 provides an overview of the service sector and information economy in Europe. More specifically it outlines, the size, location and growth of service employment on a broad industry sector basis in Europe. From this it discusses some of the conceptual problems associated with the definition and classification of service activities and concludes with some discussion of the reasons for service sector growth and the importance of the information economy. Chapter 3 then discusses some of the key economic and technological trends and issues that are occuring in services and discusses the spatial and regional development implications of such trends. These major elements include: increasing concentration of services, diversification and externalisation processes: international service transactions; moves towards deregulation, privatisation and the removal of barriers to trade in services; the role of new and small service firms; and, finally, the key change element of technological innovation itself.

Chapter 4 then goes on to look at the trends and prospects of one increasingly important segment of service economy, namely information services. As such this chapter will attempt to outline some of the main components of change that are occurring in the industrial organisation, technological innovation and locational development of this sector in Europe. Chapter 5 investigates more specifically the growth and spread of communication networks and how this has influenced the development and location of information service activity in Europe. As noted earlier Chapter 6 focuses on some of the key long term policy issues arising from the discussion and analysis provided in the report, whilst Chapter 7 provides a brief conclusion to the overall study.

2 Service industries and the information economy in Europe

The service sector, in terms of numbers of workers employed in service industries, is now the most important economic sector for all the member states of the European Community (E.C.; Table 2.1). Moreover, its importance is increasing with an estimated employment growth rate of some 15.5% within the twelve European member states (EUR.12) between 1977-83, compared with a growth in total employment of 5.1% over the same period (Table 2.2). Great care must be taken when comparing the data from Table 2.1 and too much reliance should not be placed on it, but on a broad national scale, Portugal, Ireland, Luxembourg and Italy enjoyed the highest relative growth rates over the period whilst by contrast the UK, Belgium and in particular Spain recorded low relative growth rates. In absolute terms Italy, Germany and France dominated service jobs growth accounting for over half (56%) of the employment expansion in services in the Community between 1977 and 1983.

TABLE 2.1

SERVICE EMPLOYMENT AS A PERCENTAGE OF TOTAL EMPLOYMENT
IN THE EUROPEAN COMMUNITY, 1981 AND 1984

	1981	1984
FR Germany	50.9	54.4
France	56.2	60.6
Italy	50.4	55.5
Netherlands	62.4	66.3
Belgium	62.9	68.1
Luxembourg	57.4	58.7
UK	58.2	65.5
Ireland	49.7	54.8
Denmark	63.1	66.1
Greece	42.0	45.6
Spain	45.0 (1979)	50.7 (1983)
Portugal	38.0	–
EUR.10 average (excluding Spain & Portugal)	54.4	–

Source : EUROSTAT – Regional Accounts and Commission
Services.

TABLE 2.2

SERVICE EMPLOYMENT GROWTH IN THE EUROPEAN COMMUNITY, 1977-1983
(thousands)

	1977	1983	EMPLOYMENT CHANGE 1977-1983 Absolute	Percentage
FR Germany	12,068	13,772	1,703	14.1
France	11,049	12,415	1,366	12.4
Italy	8,693	10,705	2,012	23.1
Netherlands	2,657	3,219	561	21.1
Belgium	2,017	2,158	141	7.0
Luxembourg	73	89	16	22.0
UK	13,360	14,227	867	6.5
Ireland	460	580	120	26.0
Denmark	1,302	1,547	245	18.8
Greece	1,224	1,453	229	8.7
Spain	5,113	5,195	82	1.6
Portugal	1,286	1,685	339	31.0
EUR12	58,079	67,098	9,018	15.5

Source : Data compiled from EUROSTAT Labour Force Survey,
other EUROSTAT data and for Spain, INE data (see
Appendix 1).

6

The importance of the service sector to the economy of the European Community is also supported by its contribution to total gross value added (Table 2.3), with a EUR.10 average in 1978 for services of 55 compared with 41 for industry and 4 for agriculture (see Commission of European Communities 1984b, for definition). Similarly, Green (1985, 72-4) reveals that for the Community (EUR.6) market services (covering such activities as : wholesale and retail trade; lodging and catering services; all transport and communication services; the services of insurance credit and financial institutions; and personal and business services of various kinds but excluding services provided by government; see Green 1985, 92) contributed 42.3% of gross value added in the period 1980-2 compared with only 26.7% for manufacturing. Moreover, the service share of total gross value added has been growing relative to other sectors in the economy in particular in relation to manufacturing (Table 2.4) associated with its higher annual growth rates (Table 2.5). Thus between 1973 and 1983 market services' share of total gross value added rose from 39.1 to 43.1 in the Community (EUR.7; Green, 1986).

A study by Bartells, Boonstra and Vlessand (1983) of tertiary employment in the Community found that the service sectors which experienced the highest employment growth rates in member states during the 1973-81 were: Other Services and Banking an Business Services which grew by 3.9% and 3.8% respectively (Table 2.6). For Public Administration although certain member states had high growth rates, such as Luxembourg, Belgium and Ireland, the overall growth rate was 1.9%, with the UK having the lowest growth rate of all. In Distributive Trades there was an overall employment increase of 1% although for the latter half of the survey period, 1977-1981, growth had slowed down or disappeared (apart from Italy). In Transport and Communications employment growth was a mere 0.2%, with indeed an actual decline being recorded in the latter half of the period.

TABLE 2.3

SERVICES AS A PERCENTAGE OF GROSS VALUE ADDED
IN THE EUROPEAN COMMUNITY, 1978 AND 1984

	1978[+]	1984
FR Germany	53	60.0
France	56 (1977)	59.4
Italy	51	55.7
Netherlands	61	61.1
Belgium	61	65.8
Luxembourg	54	61.4 (1982)
UK	57	57.5[*]
Ireland	49[*]	53.4
Denmark	66[*]	75.4
Greece	52[*]	53.0[*]
Spain	-	59.6[*] (1983)
Portugal	-	57.9[*] (1981)
EUR.10 average (excluding Spain & Portugal)	55	-

+   Figures are rounded.
*   Gross value added at factor cost.

Source : EUROSTAT - Regional Accounts and Commission
         Services.

TABLE 2.4

CONTRIBUTIONS OF MARKET SERVICES AND MANUFACTURING TO TOTAL GROSS[1] VALUE-ADDED 1970-72 AND 1980-82[2]

	Market Services			Manufacturing Industry			Total		
	1970-72	Change	1980-82	1970-72	Change	1980-82	1970-72	Change	1980-82
FR Germany	38.5	5.6	41.4	35.6	- 5.6	30.0	100	-	100
France	40.4	4.5	44.9	28.5	- 2.6	25.9	100	-	100
Italy	38.8	0.7	39.5	28.3	0.5	28.8	100	-	100
UK	40.1	0.9	41.0	31.2	- 6.4	24.8	100	-	100
EUR 6[3]	38.8	3.5	42.3	31.0	- 4.3	26.7	100	-	100
USA	47.1	2.5	49.6	24.7	- 3.5	21.2	100	-	100
Japan	43.5	2.7	46.2	32.8	- 4.6	28.2	100	-	100

1   Based on data at current prices.
2   To improve the estimation of changes in longer-term trends, the data are based on three year moving averages.
3   EUR6 covers the grouping FR Germany, France, Italy, United Kingdom (UK), Belgium and the Netherlands.

Source: Green (1985, 73)

TABLE 2.5

AVERAGE ANNUAL GROWTH RATES OF GROSS VALUE-ADDED AT MARKET
PRICES OVER TWO FIVE-YEAR PERIODS :
(i) 1969-71 to 1974-76; (ii) 1975-77 to 1980-82

Based on data at constant 1975 prices

	MARKET SERVICES		MANUFACTURING INDUSTRY		TOTAL	
	(i)	(ii)	(i)	(ii)	(i)	(ii)
D	3.7	4.0	2.1	1.4	2.9	2.7
F	5.4	3.7	4.9	2.0	4.3	2.6
I	4.0	3.4	4.2	3.8	3.3	2.9
UK	2.7	2.0	1.0	-2.2	2.4	0.8
EUR6	4.0	3.4	3.0	1.4	3.3	2.3
USA	3.8[1]	3.8	2.7[1]	2.1	2.8[1]	2.9
Japan	5.0[1]	5.1	5.3[1]	8.8	4.7[1]	5.6

[1] 1970-72 to 1975-77

Source : Green(1985, 74)

TABLE 2.6

TERTIARY SECTOR EMPLOYMENT GROWTH IN THE EUROPEAN COMMUNITY (EUR9) 1973–1981
(Percentages)

	Distributive Trades	Transport & Community	Banking and Business Services	Other Services	Public Administration	Total Services	Total Employment	Population
FR Germany	- 0.1	- 1.2	3.0	4.4	1.8	1.6	0.2	0.0
France	- 0.1	0.9	4.2	4.2	2.5	2.1	0.6	0.5
Netherlands	1.4	1.9	3.7	6.0	3.7	3.5	1.3	0.8
Italy	5.3	1.7	7.3	4.5	3.5	4.3	1.4	0.5
Belgium	- 0.1	1.4	3.5	4.0	5.2	2.4	0.0	0.2
UK	0.6	- 0.1	3.3	2.7	0.7	0.8	- 0.6	0.0
Luxembourg	0.5	0.2	8.6	5.7	5.8	3.3	0.1	0.4
Ireland	1.0	- 1.3	7.1	4.1	4.5	2.6	1.1	1.3
Denmark	- 1.2	- 0.4	3.6	2.6	1.9	1.2	- 0.2	0.2
EUR-9	1.0	0.2	3.8	3.9	1.9	2.0	0.3	0.3

Source : Bartells, Boonstra and Vlessant (1983, 17).

Service Location in Europe : Employment Patterns and Trends

The service sector should not only been seen as being important in terms of industrial growth and employment generation on a Community-wide or national level, but also as a key element in the economic growth and development of regions. Keeble, Owens and Thompson (1981, 133) for example in a report to the European Commission noted that "service industries are by far the most important sources of regional growth within the Community". Thus although from 1950-70 total service employment increased most rapidly in percentage and absolute terms in the central regions of the European Community and most slowly in peripheral regions (Commission of the European Communities 1984, 129) for the period 1970-80 service employment growth was highest in the peripheral regions and lowest in the central regions of the Community. This was substantiated by Keeble, Owen and Thompson (1981, 133) who found that for the period 1973-79 "service employment in peripheral regions has grown more rapidly and by a greater volume of jobs than in either the central or intermediate categories". This is also confirmed in a study by Bartels, Boonstra and Vlessert (1983, 22-3) which revealed that the highest rate of service sector employment growth between 1971-1981 was observed for peripheral regions and the lowest for central regions. Finally Keeble, Offord and Walker (1986, 93-5), in an updated analysis of regional employment trends for the period 1979-83, found that the more rapid growth of service employment in the more peripheral regions of the European Community had continued over this latest period, although because the peripheral regions have started from a low base of service employment there were only very slight changes in the aggregate distribution of services in the Community between 1979 and 1983. In overall employment terms, therefore, there was a gradual convergence between central and peripheral regions in the Community (see Figures 2.1 and 2.2 taken from Appendix 1).

Such empirical findings, in relation to the relative high growth of service employment in the peripheral regions of the Community, raises a number of explanatory possibilities, which can only be briefly raised here. The first point is that as Table 2.6 indicates, between 1973 and 1981 the peripheral EC regions had the highest annual average growth

12

FIGURE 2.1

EUROPEAN COMMUNITY ABSOLUTE CHANGE IN EMPLOYEES IN SERVICES BETWEEN 1977 AND 1983

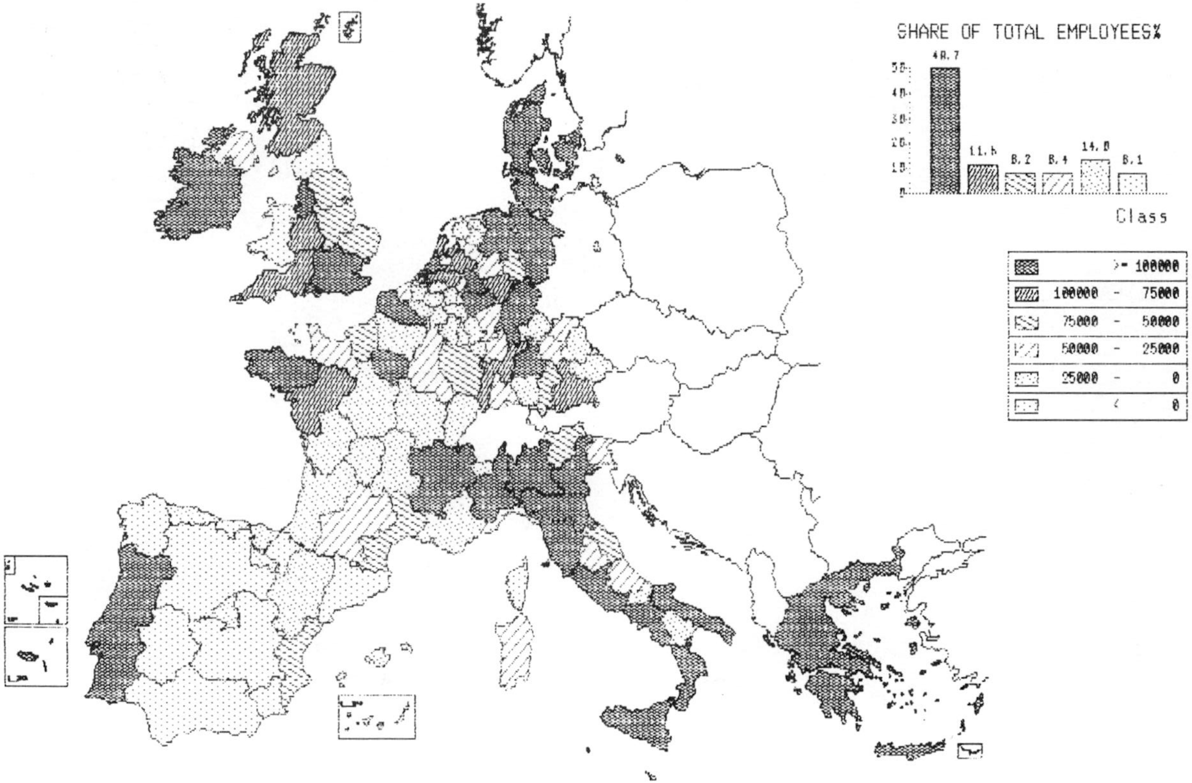

FIGURE 2.2

EUROPEAN COMMUNITY CHANGE IN SERVICE EMPLOYERS BETWEEN 1977 AND 1983 AS A PERCENTAGE OF 1977 LEVELS

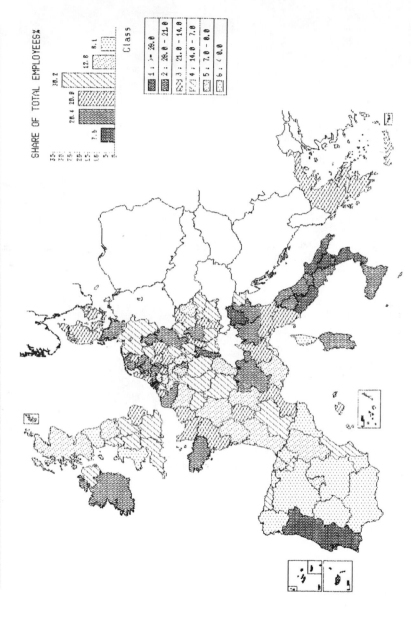

rates for population growth and total employment as well as for total services (indeed the peripheral regions had the highest growth rates in four out of five main service sub-sectors). As one would expect therefore, growth in service jobs appears to be associated (via for example, consumer spending) with a wider expansion in population and employment of the peripheral, less favoured regions of Europe. However, there are other factors which also help to explain part of this growth in service jobs. Firstly, the growth in public sector expenditure in services over this period particularly went to peripheral regions which traditionally had been underfunded in terms of public service infrastructure. Secondly, it may be that service organisations in the peripheral regions have been slower to improve labour productivity than their counterparts in more central areas of the EC and therefore labour shedding in services has been lower in these peripheral localities. More tentatively it may be that the peripheral regions benefitted from the general buoyant national economic growth, that occurred in the earlier phases of the survey period, which encouraged some convergence in growth patterns and socio-economic development. However, more research needs to be undertaken to explain this differential growth rate in service employment in the Community. Moreover, the major economic (and political) changes that have occurred in Europe since the late 1970s and early 1980s are likely to have produced substantial changes in the patterns of service employment that were observed in the 1970s.

Lastly, it is also important to emphasise the significant qualitative differences (in terms of innovativeness, export orientation and occupational type) in service employment that still exist on a regional basis. Thus although peripheral, less urbanised regions in certain European countries have benefitted from an expansion in the more dynamic, growth oriented producer and business services, there remains a marked core-periphery advantage in terms of the creation of these innovative and rapidly growing sectors. Figure 2.3 indicates that the core regions of the Community have the highest producer-consumer service ratio (see Appendix 1 for a description of this ratio) which provides an indication of a more favourable long-term service structure (see also Figure 8a, Appendix 1).

FIGURE 2.3

SERVICES STRUCTURE INDEX 1983

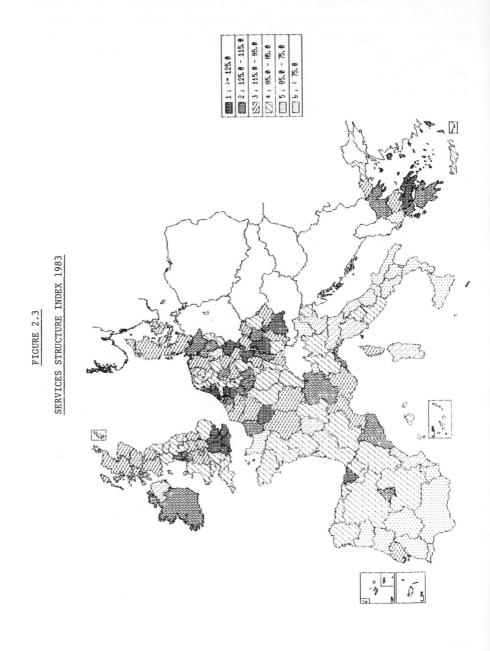

1 : >= 125.0
2 : 125.0 - 115.0
3 : 115.0 - 95.0
4 : 95.0 - 86.0
5 : 86.0 - 75.0
6 : < 75.0

16

Definition and Classification of Service Activities

In the first section basic employment trends in the service sector were revealed in aggregate terms which indicated conflating and conflicting tendencies within this heterogeneous sub-economy which makes explanation extremely difficult. It is useful therefore to look at various classification schema which may help in providing an explanatory framework that will reveal the essential elements and trends that are taking place within service activities. However, the very heterogeneity and size of the service sector has implications for its definition and classification.

Most definitions of service activity have seen it as a residual element, ie. service activity is what is left after all other activities have been excluded. A simple definition of services are those activities which do not produce or modify physical objects (commodities or products) and purchases which are immaterial, transient and produced mainly by people. However, within this definition there are vital differences to be made between service activities (Howells and Green 1986a, 2-4).

One distinction which should be recognised is between those service activities which are 'physical' in nature, involving the handling of physical goods (for example, wholesaling, retailing or distribution, and information-intensive services which primarily handle information or data, for example, research and development (R & D) or marketing; see Porat 1977). Another important categorisation relates to the two main types of functions that services perform. These are : firstly producer (or intermediate) services which provide output which is consumed or used exclusively by other industries for example market research and accountancy. Secondly, there are consumer services which provide output going directly to consumers or households, for example associated with retailing and leisure services.

A further distinction is between basic and non-basic (induced) services. It has already been noted that the traditional view of

17

service activity was that it passively followed socio-economic change in the rest of society. This was because services were largely seen as non-basic, induced activities which served purely local consumer needs, had limited multiplier effects and which simply reacted to change. By contrast basic (exportable) service activities have largely been ignored by researchers (Northern Ireland Economic Council 1982, 29-30). These latter activities can be defined as those services which are geared for national and international markets and often provide a substantial net balance of payments contribution to an area or country. In addition they are largely able to maintain self-sustaining growth independent of a particular locality and provide significant multiplier effects on a local, regional and national basis. Activities such as banking, financial and insurance services are obvious examples of service sectors which have a significant basic component; others are advertising, computer services and management consultancy.

Other key bi-polar classifications which can be made between services are _privately_ and _publicly_ provided services and, related to this, _marketed_ and _non-marketed_ services, and of importance here the _occupational mix_ of services, for example between predominantly white-collar office related services and those which have a significant blue-collar component, such as cleaning and maintenance and repair work.

In relation to this latter classification dimension, Gottman (1970) using an occupational approach divided the tertiary (service) sector into a 'quarternary' sector which involved high-order activities such as managerial, professional and high level technical staff, whilst the reduced tertiary sector covered activities such as transport and distribution and occupations associated with more routine, clerical skills. This was further refined by Abler and Adams (1977) who distinguished between tertiary (sectors providing tangible services), quarternary (covering services involving large-scale production of routine information) and quinary (relating to non-routine information production and decision-making services) service sectors.

Two additional service classification dimensions require a mention. The first is Katovzian's (1970) classification of three service groups associated with economic development :

   i) old services, in relative decline and associated with domestic and personal services;

  ii) complementary services, closely linked to manufacturing, involving business and infrastructural provisions; and

 iii) new services, associated with increasing affluence and development.

The second schema by Illeris (1985; 1986; see also Marshall, Damesick and Wood 1985) between those which are goods related, those that are associated with information processing and those services which involved distinguished persons. Goods related services cover the cleaning, maintenance and repair of goods and the transport, distribution and storage of goods. Information processing services are sub-divided into the production and combination of non-routine services and decision-making and routine information handling. Finally, services to persons covers person-related services such as education, health care and the transport of people.

What is evident from this wide range of different classificatory and conceptual schema is that they provide a different dimension and perspective (and hence classifactory framework) to delineate and categorise the same phenomena – the service sector. An important distinction to reiterate here is between those studies which examine service industries and those which seek to analyse the service component (usually assessed on an occupational basis) in all sectors of the national or regional economy (including agriculture and manufacturing). In this latter approach, research by Crum and Gudgin (1977) on a regional basis within the UK remains an important contribution. The identification and importance of the information sector in mature industrial economies however requires further elaboration and is discussed in the following section.

Service Growth and the Information Economy

The heterogeneity of the service sector is seen is illustrated in Gershuny and Miles' (1983, 28-9) attempt to categorise the causes of service sector growth. They saw such growth being related to :

(1) the increasing demand for 'intermediate' or 'producer' services from elsewhere in the economy (associated with growth arising from the 'complex of corporate activities'; Noyelle and Stanback 1984, 9).

(2) the increasing demand for services from final consumers as societies get richer (ie. services are 'income elastic'; see Gudgin 1983); and

(3) the lower rate of increase of labour productivity in the service sector relative to .manufacturing.

However, in addition to those reasons given for service sector growth in mature industrial economies must be added three others (Howells and Green 1986a, 4-5).

The first (associated with (1) above) is the process of 'externalisation' (Chapter 3) by manufacturing companies of service functions, (for example, cleaning, maintenance and public relations), which in the past were undertaken by the company and classified as 'manufacturing', and which are now subcontracted out to service firms and classified as service activities (Howells and Green 1986a, 82-95). The outward effect of this externalisation process will be a reduction in 'manufacturing' jobs and an increase in 'service' jobs although the overall net effect for the economy is often no change or indeed a slight employment reduction as service firms provide the function more efficiently.

The second aspect of service growth is the increasing internationalisation of service activity. The service sector is still heavily regulated and restricted in relation to foreign trade; nevertheless certain sectors in Europe have provided a valuable contribution to national balance of payments. With the rapid annual growth in the business service market in the largest industrial countries, the growth potential for European service companies which are able to compete on an international scale in these fields would appear to be considerable.

The final component relating to service sector growth is the actual creation of new service activity. Again problems of definition and measurement occur with trying to discover what services are totally 'new' and in what degree other new services, such as video shops/libraries, replace existing service activities, such as theatres and cinemas. However, some new activities, often associated with new forms of technology, such as on-line information services, have led to 'real' service growth. New and existing services being provided via new kinds of distributed system or a substantially altered format also have important employment implications within services.

There has however been discussions on the relative strengths of these service sector growth factors. The first revolves around the importance of the growth of intermediate demand by industry and commerce for producer services as compared with final demand by consumers and households. Evidence from work by Green (1985, 80-9) suggests that the link between increasing living standards (GDP per head) and increased overall consumption of services does not relate so much to the increased private consumption of services as suggested by Gershuny and Miles (1983; although it should be noted that they suggested that the 'self-service economy would reduce the consumption of services by private consumers and households') but instead owes more to the rapid volume growth of the consumption of services by industry (Green 1985, 87). Thus over the period 1975-82 (Table 2.7) more than half of the growth of market services output is accounted for by the growth of intermediate consumption by industry and by services, whilst less than one-third of the growth of output is accounted for by the growth of household consumption of services (Green 1985, 88). Clearly the increasing growth and complexity of demand for services by manufacturing and service companies provided both internally (Williamson 1975; Rugman 1981; 1982; Radner 1986; Taylor and Thrift 1986) and externally (Howells and Green 1986a; Howells 1987a) is of key importance in the development of the service economy in Europe. In particular the growth of information related service activities is intimately linked to the growing sophistication and information content of industry as a whole (Rada 1984, 7).

TABLE 2.7

THE GROWTH OF SERVICES OUTPUT IN THE EUROPEAN COMMUNITY (EUR6, 1975-82)
: THE BALANCE BETWEEN DEMAND AND OUTPUT

	Components of demand for services				
	Intermediate Consumption by Industry	Intermediate Consumption by Services	Final Consumption of Households	Exports of Services	Total
Weights from input-output for 1975	0.192	0.345	0.387	0.076	1.000
Estimated volume growth rates[1] (%)	3.6	3.1	2.4	6.9	3.2
Contribution to growth of services activity output[2] (%)	0.69	1.07	0.93	0.52	3.2

1 Estimated average annual growth rates 1975-82
2 There is a slight discrepancy between the sum of components and the total given due to rounding

Source: Green (1985, 89)

Another issue is that labour increases in services are now rapidly catching up, or indeed overtaking, the manufacturing sector since the late 1970s in Europe. Thus on a European level Green (1985, 75-6) shows that the gap in average annual growth rate in labour productivity between manufacturing and market service sectors in the European Community (EUR6) has narrowed over the 1970s and early 1980s (Chapter 3).

Clearly all these changes in the significance of various factors associated with the growth of services in output and employment terms will have important regional development implications (Chapter 3). These growth factors also suggest why attention has been increasingly focussed upon the role of the information processing as a major growth component in the post-industrial society (Bell 1974). The foundation for this shift stems from a study by Machlup (1962; see also Rubin and Huber 1986) which stressed the major importance of information and knowledge in the modern economy. More fundamentally this work was extended and operationalised by Porat (1977) who produced an input-output analysis of the contribution of the information sector of the US economy. Porat (1977, 4) identifies two main information sectors, these were :

(1) The primary information sector which includes those firms which supply the information goods and services exchanged in a market context.

(2) The secondary information sector which covers all these information services produced for internal consumption by government and non-information firms.

Porat (1977, 106-7) also makes a distinction between information and non-information occupations. Within information occupations there are three major classes :

(1) Those occupations involved with 'markets for information' which includes workers whose output or primary activity is an information product. This group is further divided into knowledge producers and distributors.

(2) Secondly, there are those occupations which provide information in markets where their output is not knowledge for sale, but rather serve as information gatherers and disseminators. These workers

move information within firms and markets and include market search and co-ordination specialists and information processors.

(3)   Lastly there are information infrastructure makers whose occupation involves operation of information technology equipment to support the previous two activities.

Table 2.8 provides a compressed classification of Porat's information sector and occupation groups based on OECD's (1981) operationalising his system using ISIC and ISCO categories for their international study of the information economy (Miles 1985, 11-12).

Porat's (1977) study revealed that the information sector of the US economy accounted for over 40% of the total workforce and 46% of the country's GNP in 1967 (see also Rubin and Taylor 1981). A study for the Canadian government revealed that the information sector increased its share of total employment between 1961-1971 from 35.8% to 39.7% (Canadian Economic Services Limited 1977), whilst the employment share of the information sector in 1971 in Australia was 27.3% (Lamberton 1977; 1983). This growth in the information sector in advanced industrial economies has been confirmed in a large scale study undertaken by the OECD (1981), covering nine countries including three from the European Community : France, Germany and the UK. The study indicated that in all the countries examined there was a progressive shift towards those occupations concerned with the creation and handling of information (Figure 2.4). On average, over the range of periods and countries studied, for each five year period the information sector gained an extra 2.8% share of total economically active workers. The greatest growth came from the information processing occupations (nearly 59% of the observed growth rates), followed by information producers (21% of overall growth rates), information distributors (13%) and lastly infrastructure workers (7%; OECD 1981, 23). In a recent update of their 1981 analysis the OECD discovered that there was a slight reduction in the rate at which information occupations grew in advanced industrial economies (OECD 1985a 3). Thus for the latest five year periods available information occupations gained an extra 2.3% in share of total economically active compared with the 2.8% relative growth noted earlier. More specifically however a study in Britain has indicated that the proportion of information workers in 1981 exceeded 45% (Hepworth, Gillespie and Green 1987, 796) compared with the OECD estimates for UK as a whole of 26.7% in 1951 and 35.6% in 1971. Other studies have stressed the importance of the information sector on a regional level in Ireland (Bannon and Blair 1985) in the US (Schement, Leivoroun and Dordick 1983) and Australia (Mandeville et al. 1983; see also Mandeville 1985).

Nevertheless, 'information society' analysis has come under criticism. Apart from the criticism of the crude classification of the information sectors and occupations it has been suggested that it also ignores some of the other key dimensions of the service economy. Information society theory, Miles (1985, 13) notes have added an extra dimension in which to view the service economy and has provided useful analysis of information sources, but it has provided little substantial analysis or insight of specific trends in quite heterogeneous activities.

Overall, however, the importance of growth and actual phenomenon of the information economy in the European Community needs to be acknowledged in any study analysing the service sector in Europe, particularly in relation to future trends and developments. For this reason Chapter 4 undertakes a more specific analysis of information services industries and the implications for their growth and evolution on European regional development, whilst Chapter Seven presents policy issues centred on the information services and economic development.

TABLE 2.8

(A) Inventory of the Information Sector

ISIC Division	Components of Primary Information Sector
1 (Agriculture, Forestry, Fishing)	part: 1120, 1302
2 (Mining, Quarrying)	part: 2100, 2200
3 (Manufacturing)	
32 (Textile, Wearing Apparel, Leather Goods)	part: 3211, 3233
33 (Wood Products)	part: 3319, 3320
34 (Paper Products, Printing, Publishing)	part: 3411, 3412, 3419 all: 3420
35 (Chemical, Coal, Petroleum, Rubber, Plastic Products)	part: 3529, 3560
36 (Non-metallic Mineral Products)	part: 3620
38 (Fabricated Metal Products, Machinery, Equipment)	part: 3811, 3812, 3819, 3822, 3824, 3831, 3839 all: 3825, 3832, 3852, 3853
39 (Other Manufacturing)	part: 3909
4 (Electricity, Gas, Water)	No information components
5 (Construction)	part: 5000
6 (Wholesale and Retail Trade)	part: 6100, 6200
7 (Transport, Storage, Communication)	part: 7111, 7132, 7191 all: 7200
8 (Finance, Insurance, Real Estate, Business Services)	part: 8101, 8200, 8310, 8330 all: 8102, 8103, 8321-5, 8329
9 (Community, Social and Personal Services)	part: 9331, 9512, 9514, 9519, 9530 all: 9310, 9320, 9411-5, 9420, 9492

TABLE 2.8 (Continued)

(B)   Information Occupations

Category	ISCO Code Numbers

Information Producers

Scientific and Technical   0-11, 12, 13, 22, 23, 24, 26, 27, 28(part), 29, 51, 52, 53, 81, 82, 90; 1-92

Market Search and Coordination Specialists
4-10.20, 22, 31, 41, 42, 43-20

Information Gatherers   0-18.30, 31, 33.20; 1-20.50; 3-59.30, 59-45, 91.50; 4-43.30; 5-89.20; 7-54.70; 8-59.20; 9-49.90

Consultative Services   0-21, 32, 61, 69, 75.20, 83, 84.20; 1-10(part), 21, 29, 39.20, 62

Information Producers nec   1-51.20, 71.20

Information Processors

Administrative and Managerial   1-22, 39.40; 2-01, 02, 11, 12, 19; 3-10; 4.00

Process Control and Supervisory   0-33.40, 41.40, 42.30; 3-0, 5, 91.20; 4-21; 5-20, 31.20; 6.00.30, 32.20; 7-0

Clerical and Related   1-10.20; 3-21, 31.10, 31.20, 39.20, 39.30, 39.40, 91.30, 92.20, 92.30, 93, 94, 95, 99.20, 99.30, 99.40

Information Distributors

Educators   1-31, 32, 33, 34, .35

Communication Workers   1-51.30, 59(part), 73.30, 73.40, 73.50, 79.20

Information Infrastructure Occupations

Information Machine Workers   1-63; 3-21.50; 22, 41, 42, 99.50; 8-49.65, 62; 9-21, 22, 23, 24, 25, 26, 27

Postal and Telecommunications   3-70, 80; 8-54, 56, 57.40, 61

Source : Miles, (1985, pp.11-12).

FIGURE 2.4

CHANGES IN THE SHARE OF INFORMATION OCCUPATIONS
IN ALL ECONOMICALLY ACTIVE OVER THE POST—WAR PERIOD

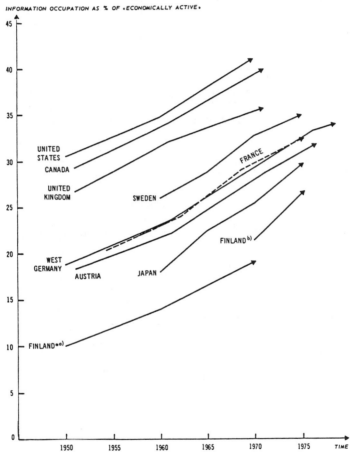

INFORMATION OCCUPATION AS % OF «ECONOMICALLY ACTIVE»

* Data for Finland was derived from two separate sources : (a) I. Pietarinen ; (b) The Central
  Statistical Office of Finland, both sources using a rather more restricted definition of
  «information occupations» than that of Table A I. Absolute values for any given year are,
  therefore, not strictly comparable with other countries, although the trend is still of interest.

Source : OECD (1981, 25).

3 Economic, technological and locational trends in European services

Introduction

Chapter 2 provided a short description on the various ways services have been defined and classified. However, although these different classifications are useful on a conceptual level, in practice they are difficult to apply on an empirical level. More fundamentally however such classificatory schemas are firmly embedded in a macro-economic, aggregate view of services, providing very much a 'top-down' approach and supplying little or no explanatory input, particularly on a dynamic level. As such there is no indication of : wider trends or phenomenon that are occurring in the sector; the nature and role of the key agents and organisations operating in services; the impact of more structural trends (such as acquisition or diversification) that are occurring in service industries; or the role of external pressures on the sector (associated with, for example, the opening up of inter-national trade or government deregulation).

It is because of these important issues that the rest of this study, and this chapter in particular, will adapt a micro-economic level view focussing on the key structural, technological and organisational issues that are occurring in services in the European Community. From these emerging or prospective trends certain regional development implications will be outlined.

Service Concentration, Linkage Networks and Location

A key element in the evolution of the service sector in Europe and its locational development is that of increasing concentration of ownership. This is not to suggest that concentration ratios are the same everywhere within the European Community or that rises are going on at the same rate, nevertheless increasing concentration in services in Europe is occurring steadily over time.

The increasing concentration in European service industries has been confirmed by a number of sectoral studies. A study by Bartells, Werkhoven and de Kruijk (1984, 62-3) on retailing employment found that there was a process of concentration going on, with larger enterprises taking an increasing share of retail service jobs over time. Thus between 1974 and 1978 they found that in six EC countries the largest enterprises increased their employment share in all countries except Belgium, with the fastest rates of increase occurring in Denmark and West Germany. In relation to banking a recent survey of European banks by consultants Arthur Anderson (1986) revealed that two-thirds of the European bankers they interviewed predicted that over the next decade larger institutions would expand further at the expense of smaller ones, especially investment banks and savings institutes. Work by Aaronovitch and Samson (1985, 67) on the insurance industry in the countries of the EC revealed that the insurance sectors in all EC member states were highly concentrated (Table 3.1). Moreover, in most countries the distribution of firm and group sizes was highly skewed within the cluster of dominant groups at the top of each sector. However, the study found that in general the high overall level of concentration did not significantly change in the period 1975-1980. Finally as one would

TABLE 3.1

CONCENTRATION RATIOS BASED ON PREMIUM INCOME[a] FOR THE WHOLE INSURANCE SECTOR BY COUNTRY IN 1980[b]

	CR1 %	CR2 %	CR3 %	CR4 %	CR5 %	CR7 %	CR8 %	CR9 %	CR10 %	CR11 %	CR12 %	CR15 %	CR20 %	Oligopoly 'Arena'	'Axis'
Belgium[a]		30g	36d						58				75		
Denmark					41						67			8-10	2-3
France				38g			55g				66g	70+g		4 life / 20 non-L	2 / 2
Germany	15c					45g									
Greece	20		49				62				70			3-4	2
														Bank controlled	
Ireland			32e	51g							54c				
Italy	22g	46g			57g	67g		73g		77g					
Netherlands	23g	34g							70g						
UK[e]					30g				51g			62g			

Source : Aaronovitch and Sawyer (1985, 68)

Notes :
a The basis of calculation for the concentration ratios for Belgium is Total Income from Premiums and Investments.
b 1979 for Belgium, Denmark, France, Netherlands and the UK.
c Konzern under unitary direction (Germany); single company level (Italy).
d These three largest groups are controlled by Societe Generale de Belgique.
e Excludes Lloyd's.
g Group level of companies.

expect, the dominance of large organisations in the service sector is also reflected in the proportion of service employees working for large establishments (Figure 3.1). This employment pattern is discussed in more detail later in the context of the 'dual economy' model.

This sectoral concentration of ownership in service industries is also apparent in the geographical concentration of ownership and control. Previous studies have demonstrated the concentration of industrial headquarters in the UK (Parsons 1972; Evans 1973; Westaway 1974a; 1974b; Goddard and Smith 1978), the USA (Armstrong 1972; 1979; Semple 1973; 1977; Pred 1974; 1977; Burns 1977; Borchett 1978; Daniels 1985) and Canada (Phillips 1982) and have indicated the preference of headquarter units for large metropolitan areas (Dunning and Norman 1979). These studies link in with wider discussions associated with the spatial concentration of strategic, orientation service functions within large multi-unit organisations (Thorngren 1970; Goddard 1975; Pred 1977; Lazzeri 1986).

These findings are borne out in a European-wide context in an analysis of the headquarters location of a 100 giant service corporations in Europe. Thus sectoral concentration in service industries is also reflected in the geographical concentration of ownership and control in terms of the headquarters location of these giant service corporations (see later). This is evident in Table 3.2 which indicates the Head Office (HO) location of a 100 major service in Europe (the largest twenty companies in the following sectors : banking, insurance, transport, advertising, hotels and restaurants). What is apparent is that three major cities : London, Paris and Frankfurt, have nearly half of all the HOs of these major service corporations, with London alone having a quarter of all the HOs located there. Geographical concentration of such strategic planning and decision-making functions in these cities has major implications for the demand for other intermediate business and producer service activities in terms of volume and, more particularly, quality (for a review of the UK situation, see Howells and Green 1986a, 141-54).

A number of studies have shown that headquarter units will centralise a high proportion of corporate intermediate service demand linkages within their local area or region. This demand will be satisfied either in-house in offices at or near the headquarters site, or externally via suppliers located nearby. Thus a high proportion of intermediate service inputs required by externally controlled manufacturing and service establishments in peripheral regions will be supplied both on an internal and external basis, from outside the region (see, for example, Marshall 1982a; 1982b). This position is further entrenched via takeover and merger activity whereby centrally located large multi-establishment service corporations acquire smaller indigenous service companies established in peripheral, less favoured regions. Merger and acquisition amongst UK insurance companies, for example, has created larger, more centralised, organisations in which internal provision of accounting, marketing and computer services, via head offices, has replaced the autonomy of smaller, often locally-based insurance companies which obtain producer services from local suppliers (Daniels 1986a, 308). Moreover, because of the sophisticated and specialised requirements of headquarters functions a highly concentrated pattern of demand for, and supply of, more specialised and innovative service activities around major metropolitan areas of Europe will also be encouraged. A good example of this (Table 3.3) is the close spatial association between UK computer service users and providers (in terms of their office location).

However, on a national basis Table 3.2 tentatively indicates that some countries are very much more highly concentrated in terms of headquarters functions of these top service companies than others. Thus of the top six countries ranked in terms of headquarters numbers, the UK, Sweden and, to a lesser extent, France are very much more highly concentrated in terms of capital city domination than the federal government structures of Germany, Switzerland and in some senses the Netherlands. The structure and functioning of government may have certain implications therefore for the centralisation or otherwise of industrial service corporations in Europe which needs to be considered. Moreover, although Table 3.2 provides only a static picture of the

TABLE 3.2

HEADQUARTERS LOCATION OF A 100 MAJOR SERVICE CORPORATIONS IN EUROPE, 1983

		NUMBER OF HEADQUARTERS FOR :		
Rank	Name of City	a) City	b) Total for country including the leading city	a) as a % of b)
1	London	25	27 (UK)	93
2	Paris	14	20 (France)	70
3	Frankfurt	9	17 (Germany)	53
4	Zurich	6	15 (Switzerland)	40
5	Stockholm	5	6 (Sweden)	83
6	Amsterdam	2	5 (Netherlands)	40
7	Rome	2	3 (Italy)	67
8=	Brussels	2	2 (Belgium)	100
8=	Oslo	2	2 (Norway)	100
10=	Madrid	1	1 (Spain)	100
10=	Helsinki	1	1 (Finland)	100
10=	Copenhagen	1	1 (Denmark)	100

Includes 100 leading corporations (includes some public corporations and major subsidiaries) who are the top 20 European service companies in the following sectors :

    1) Banking   (ranked by assets $)
    2) Insurance   (ranked by premium income $)
    3) Transport   (ranked by sales $)
    4) Advertising (ranked by turnover $)
    5) Hotels and Restaurants (ranked by sales $)[+]

[+] data modified.

Source: compiled from data from Europes Largest Companies Limited (1985).

TABLE 3.3

REGIONAL DISTRIBUTION OF UK COMPUTER SERVICE OFFICES AND USERS

LOCATION	% OFFICES	% USERS
Greater London	32.0	21.3
Rest South East	24.1	24.1
Total	56.0	45.4
West Midlands	8.0	8.3
North West	9.3	10.4
Wales	1.8	3.7
Northern Ireland	0.7	1.1
South West	5.8	5.3
Scotland	4.9	8.3
North and Yorkshire & Humberside	7.9	9.6
East Midlands and East Anglia	5.6	8.0
Total Provinces	44.0	54.6
TOTALS	100.0	100.0

Source : Gillespie et al. (1984), 97) using data from IDC Europa.

spatial concentration of giant service corporations in Europe, evidence from the UK, at least using data from Goddard and Smith (1978), suggests that the spatial concentration headquarters of large non-manufacturing, mainly service, corporations was increasing during the 1970s and indeed more rapidly than for head offices of large manufacturing companies (Howells and Green 1986a, 142).

Obviously a great deal more research needs to be undertaken but on a regional development basis the increasing spatial concentration of ownership and control in services is of concern here. The 'branch plant economy' syndrome evident in peripheral regions in terms of manufacturing (relating to a situation whereby a large amount of ownership and control of enterprises in key sectors lies outside the region concerned; Firn 1975, 394) would appear, from limited UK evidence (Howells and Green 1986a, 144-5), to becoming increasingly applicable in the service sector as overall concentration levels rise and there is a continuing centralisation of control to the core regions. As such, service establishments in these more peripheral areas will become associated with the delivery of less sophisticated, routine services, with high-order, generative functions being centralised in larger, core metropolitan regions. Support for this view is indeed provided by Cuadrado's (1986) detailed study of the Comunidad Valenciana region of Spain where he found that externally controlled service establishments were oriented towards the more strategic or specialised service sectors, such as finance, insurance, accountancy and management consultancy, market research and information systems. He also found that the head offices performed the more specialised and complex service operations for their branch establishments located in the region, noting that "the head offices usually perform the logistic and organisational tasks, as well as the direct provision of those more complex and sophisticated services" (Cuadrado 1986, 127).

Diversification and Acquisition in Services : Related to the trend of increasing concentration in services has been the process of diversification and acquisition. There are three strands in the diversification process in relation to services (Table 3.4.I). Firstly, major manufacturing companies in Europe have diversified into the service sector either via acquisition or the development and marketing of their own internal service activities (such as computing services) through the processes of internalisation and 'externalisation'. Similarly service companies are diversifying into other service sectors, associated with the emergence of new service activities and government deregulation, particularly within the banking and financial sectors. Finally, some service companies have diversified into manufacturing activities, perhaps the most common example of this are retailers starting to manufacture their own products. These three elements in the diversification process has meant that the distinction between the service sector and the rest of the economy has become increasingly blurred. Moreover, companies operating in the service economy are no longer confined to a single service sector, but cover a range of service manufacturing and trading operations forming giant multinational service or industrial conglomerates (Clairmonte and Cavanagh 1984, 216-7).

A wide range of examples of this diversification process in UK services has been provided (Howells and Green 1986a; 1988), and it would appear that this phenomenon has been going on for a longer time in the UK. Nevertheless there is a growing number of examples in other Community countries (Table 3.5). Thus Montedison, the Italian chemical manufacturing company, which owns Standa - a large retailing chain in Italy, has taken over Bi-Invest a large financial and industrial group which in turn has a controlling stake in La Foundiara, one of Italy's largest insurers. From this it is now offering pension fund services, using the Standa retail subsidiary as its base for a wide variety of financial services. Similarly Benetton, the Italian clothing company, has a long term strategy to diversify into financial services. Benetton already has two financial services subsidiaries, Incapital and Infactor;

TABLE 3.4

DIVERSIFICATION AND EXTERNALISATION OF SERVICE OPERATIONS :
CLASSIFICATION MATRIX AND MODEL

I : Matrix

Main operational sector of company :	New sector diversifying into :
1. Manufacturing	Service
2. Service	Service
3. Service	Manufacturing

II : Model

Diversification accomplished via :

A   Acquisition/takeover of an existing company in the new sector.
B   Internal developments and externalisation of service operations within companies :

 i) Creation of new autonomous department/function within the company.

 ii) Formation of subsidiary company where parent company has controlling share.

iii) Formation of subsidiary or associate company where 'parent' company has minority ownership.

B i) to iii) represents a possible progression related to the internal development of a separate enterprise within the main company which in turn may lead to :

 iv) The complete spin-off with the disposal (sale) of the company or a management buy-out associated with the final stage of the externalisation process.

Source : Howells and Green (1986a, 83; 1988).

has bought 10% share holdings in leasing companies in France and Germany, and has acquired a 35% stake in an Italian insurance company controlled by Prudential, a leading UK insurance and financial services company. It has also established a financial capital joint venture company with GFT, another Italian clothing manufacturer, which provides corporate finance and related investment banking activities, syndicated loans and currency swaps. Benetton and GFT now have cross shareholding links in their financial service subsidiaries with GFT owning 25% respectively of Incapital and Infactor, whilst Benetton now has a 25% stake in P&I an insurance brokers owned by GFT. Indeed it is in financial services that most manufacturing and service companies have sought to diversify into (Table 3.5). In part this has been a natural progression from their existing operations, such as Marks and Spencer with its chargecard activities, but it also relates to a desire to provide a more comprehensive financial service package to existing customers (gaining scale economies through greater utilisation of branch networks) and above all from the good returns made from financial service operations overall.

A key means by which company growth (and hence ownership concentration in services) and diversification can be achieved is via acquisition and takeover (Table 3.4.II). In most European countries acquisition and takeover in services has occurred much later than in manufacturing and it has been only in the last ten or fifteen years that this has become an important element in the development of the European service sector. More recently it has been encouraged by deregulation moves in certain EC countries which has allowed banking, insurance and other financial and business markets to be opened to competition from organisations traditionally excluded from such markets. Indeed in terms of service diversification perhaps the best example of European service companies diversifying into other service sectors are banks. Thus a number of EC banks (Deutsche Bank, Paribas Banques, Bruxelles Lambert and Banque Belge) are diversifying into other financial activities via acquiring UK stock exchange firms operating in international markets associated with the deregulation of the UK financial market (the 'Big Bang') in October 1986 (Howells and Green 1986a, 117; 1987).

TABLE 3.5

EXAMPLES OF SERVICE DIVERSIFICATION IN EUROPE

DIVERSIFICATION TYPE (Table 3.4)	NAME OF COMPANY	NATIONALITY	MAIN OPERATIONAL SECTOR OF COMPANY	NAME OF DIVERSIFIED UNIT OR ENTERPRISE
1A/B	Montedison	Italian	Chemical Manufacturer	Standa
1A				Bi Invest
1A				La Fondiavia
1A/B	Benetton	Italian	Clothing Manufacturer/ retailing	—
				—
1	Air France	France	Airline	Meridien
1/2A	Societe Generale de Belgique	Belgium	Industrial Conglomerate	Tanks Consolidated
1/2A				Dillan Read
1/2A				Mercapital
2A	Aachever und Muechener	German	Insurance	Bank fuer Gemeinwirtschaft
2A	Allianz	German	Insurance	Bayevische Hypotheken-sud-Weohsel
2Bii)	Marks and Spencer PLC	UK	Retail	St Michael Financial services Limited

40

SECTOR OF DIVERSIFICATION	OTHER COMMENTS
Retailing	
Financial services	
Insurance	Currently controls 40.5% of the company's shares
Financial services and venture capital	Cross shareholding links with GFT in : Incapital, Infactor and P & I
Factoring and leasing services	
Hotels	Owns 50 hotels worldwide
Banking	Acquisition of UK Merchant Bank
Financial services	50% stake in international subsidiary based in UK of US investment bank
Financial services	25% stake in Spanish Investment Bank
Banking	Proposed takeover for 51% of shares
Banking	Part control
Banking	Licensed deposit taker; soon to have full banking license. Has 1.3 million chargecard holders.

The above are also represent acquisition activity on an extra-European
scale. The top French computer service companies, Cap-Gemini-Sogeti,
SG2 and CiSi, also provide examples of acquisitive strategies in both
Europe and the US which are pursued to fulfill policies of developing
geographical expansion and sector specialisation. Interestingly their
acquisition programmes only began since the late 1970s. Another example
are the two major acquisitions by Allianz, a leading German insurance
company, of Cornhill Insurance in the UK and Riunione Adratica di'
Sicurta of Italy. Similarly Societe Generale de Belgique, Belgium's
giant industrial and financial holding company has also adopted a
strategy of developing an international financial services operation
through its acquisition or part control of UK, US and Spanish financial
service companies (Table 3.5). Moreover, Societe Generale de Belgique
has close links with three insurance companies A.G., Assubel and Royale
Belge, which takes the form of complex shareholding links and
interlocking directorships forming a network of reciprocal
participation. It also has, via other companies such as SOFINA, links
with finance companies (for example, IPPA). However, one of the severe
limitations of pan-European acquisition policies remain national
government restrictions on foreign ownership in a whole range of service
sectors.

Internalisation and Externalisation of Services : Related to the process
of diversification has been an important phenomenon, leading to the
emergence and growth of new corporate actors in services in the UK and
other EC countries, that of internalisation economies (Williamson 1975,
29-30; see also Rugman 1981; 1982; Radner 1986; Taylor and Thrift 1986)
and the 'externalisation' process (Howells and Green 1986a; Howells
1987a). Externalisation is in turn related to the commodification and
emergence of new service activities within large companies, particularly
in relation to new information services (Chapter 5). The discussion
here of the externalisation process is a revised and more refined
version of what was presented earlier in Howells and Green (1986a,
82-95; 1986b, 125-131).

Externalisation can be viewed on two different levels. On one
level it involves the externalisation of services by firms, involving
the closure of in-house service functions and the purchase of service
products from other companies. At the other level is the process
associated with the emergence and development of service operations
within companies producing services initially for in-house consumption
only but which subsequently come to sell their services to other
companies. This latter process may lead to the eventual hiving off or
complete disposal of the service enterprise by the present company.
These two separate but closely related externalisation processes will be
discussed in turn.

Externalisation of services at one level has been taken to be the
process whereby companies decided to no longer provide a particular
service product in-house, such as cleaning, maintenance or public
relations, but instead to purchase it from another company. This form
of externalisation therefore revolves around whether a company should
produce its own services or purchase it from specialist service
companies or operations. The degree to which a service function is
internalised or externalised depends on a variety of factors, chief of
which are the size of the company (with large firms more likely to
internalise service requirements) and the type of service being
considered (Marshall 1980; 1982). Some functions such as legal services
have generally had a much higher level of externalisation than others,
for example accounting, regardless of firm size (Marshall 1982, 24).
Nevertheless it is clear that certain service functions are being
increasingly externalised, even by large corporations. It is possible
that this trend in information services is being encouraged by the
growth of inter-organisational computer-communication networks, which
allows certain information and data services to be more easily and
cheaply provided in an inter-organisational basis (see Estrin 1985,
281-2). Service functions which are becoming more likely to be
purchased externally cover such activities as cleaning, maintenance,
security, distribution and public relations.

There are a number of basic motives for externalising services. These are associated with potential cost savings, the attraction of better quality service and the increasing technical complexity or specialisation of functions which the firm feels it can no longer afford to maintain. The final motive is associated with the other form of internalisation/externalisation processes relating to the disposal of a service operation which is selling service products to other companies. The company will receive cash proceeds from the sale of the operation as an on-going concern (Howells and Green 1986a, 85). However, of crucial importance in the whole decision of whether to continue to maintain an in-house operation or to purchase services externally is the actual strategy of the company. Some companies are happy to maintain in-house service provision even when they are less efficient for non-cost factors or because it forms part of an active strategy to encourage the development of service operation to sell its services to other companies. Other companies however are seeking to slim down their operations to focus on their core in-house service activities in order to improve efficiency and flexibility (Atkinson 1985) which in turn is often as part of an overall disposal package which aims to concentrate the firm's activities on selected sectors where it intends to specialise.

An important factor often inhibiting service externalisation is the issue of confidentiality which is particularly important in certain functions such as computing and data processing, legal and public relation services. Moreover, the main motive for externalisation is only in some cases that of cost savings. Often it is that of the increasing technical complexity or specialisation of the function. Thus, for example, the role of security has become more sophisticated and complex over the last twenty years. Public relations is a relatively new function for most companies and frequently it is an activity which companies have tried to do internally but eventually found it more satisfactory if the task was handled by a specialised public relations agency. Maintenance was an activity which most service companies externalised from the start if the equipment was at all complex. On this basis currently a much higher proportion of maintenance tasks in the UK are externalised by service companies compared with manufacturing (Marshall 1982, 24). Over time this gap appears to be reducing as specialised maintenance companies, from their established market base in the service sector, are now beginning to offer competitive rates, in particular for large or one-off maintenance problems in production which manufacturing companies find hard to plan for.

Service externalisation associated with the emergence and development of service operations or enterprises within firms which begin to sell their services to other companies can be conceptualised in a number of stages and reference is made back to Table 3.4(II). The first stage (Bi) is associated with a department, unit or team within a company producing services for in-house consumption which starts to sell its services to other companies. A key initiating factor here may be a reassessment of the costing and valuation of the services it produces in-house. More stricter costing schemes may be introduced which may in turn lead to a pricing policy being adopted. This may reveal that the service can be bought more cheaply from outside, however it may also indicate that the company can, if it gains sufficient outside demand, make a substantial profit by selling its services to other companies. In this latter case services generated for internal consumption subsequently become commodified and developed into a service product. A UK example of this is ICI's 'Assassin' computer software package that the company's data processing/computer services department produced and used and which was then found to be marketable to other companies (sales are now being handled by ICL). As the external market for a particular service expands the department may be given more autonomy and become a profit centre in its own right. At some stage if the market for such service continues to grow, the company may consider setting up a wholly or partly owned subsidiary or associate company (Stages Bii) and iii)), so that the service can be more effectively marketed and expanded. As noted above some companies may go so far as to sell off (Stage Biv)) these externalised service operations. Thus the giant Anglo-Dutch food company, Unilever, decided to sell-off the UCSL with its 375 staff for £16 million in 1984. Interestingly, it was sold to EDS a US-based computer systems integration and management company which was acquired by General Motors for some $2.5 billion earlier in 1984. The actual shedding or disposal of a service by a company (Biv)) should however be best conceived of as the final stage of a much wider process of the externalisation, commodification and emergence of service activities. In particular it should be stressed that stages Bi) - Biv) are not meant to imply that there is a straight progression from one stage to another. Many companies continue to operate often large units, selling services to other firms, in an embryonic enterprise-form closely integrated within the company's organisation.

Indeed the computer service sector provides a good example of this whole process on a European as well as UK basis (Howells 1987b). A significant number of major computer service operators in the UK market have arisen out of in-house computing service departments within large multinational business corporations. Thus the six largest computer service operations in 1985 in the UK are part of large multinational corporations with, for example, Scicon International owned by British Petroleum and Centre File owned by National Westminster Bank. This is also true of four key French computer service companies noted earlier : Cap-Gemini-Sogeti, CiSi, SG2 and GSi, who after IBM-INS were the four largest computer service companies operating in Western Europe in 1984 (ranked respectively 2, 3, 4 and 5). Thus GSi and Groupe CiSi grew out of internal computing service bureaux of two large industrial concerns, Compagnie Generale d'Electricite (CGE) and Commissariat a l'Energie Atomique (CEA) respectively (Table 3.6) and later given their own autonomy. Similarly Groupe SG2 were formed under the auspices of the Banque Societe Generale, whilst Cap-Gemini-Sogeti arose out of a merger of three separate software houses in 1975.

Some of the key computer service and software operations in Europe have therefore arisen out of internal computer service departments of large multinational corporations which then become subsidiaries, divisions or profit centres embedded within the corporation, or sometimes fully externalised through a sell-off or management buy-outs. As such a number of large computer service operations remain under the

46

TABLE 3.6

FOUR LARGEST EUROPEAN-OWNED COMPUTING SERVICE COMPANIES IN WESTERN EUROPE

| COMPANY | OWNERSHIP/ PARENT COMPANY | DATE ESTABLISHED | ORIGINS | SIZE (Employees, 1985) | | | TOTAL TURNOVER 1984 ($ million) |
				1. France	2. Rest of World	3. Total	
Cap-Gemini-Sogeti	Management/CGIP	1975	Merger of three companies : CAP (est. 1962) SOGETI (est. 1967) GEMINI (est. 1968)	2,301	2,233	4,534	207.7
Groupe CISI	Commissioriat a l'Energie Atomique (CEA)/ Banque National de Paris (BNP)	1972	Started as internal computer service operation of CEA	2,000	1,000	3,000	172.7
Groupe SG2	Banque Societe Generale	1970	Started by Banque Societe Generale	3,500	700	4,200	150.6
GSI	CIT Alcatel (controlled CGE)	1971	Started as internal computer service operation to Compagnie Generale d'Electricite (CGE)	1,422	926	2,348	135.9

Source : Howells (1987b).

control of their parent company and given various degrees of autonomy and status within the parent group. In most cases however, they still provide an internal service to their parent establishment even if it is often under a market environment. Interestingly Denmark has a number of significant computer service companies controlled by common interest groups or with a minimum of 50% captive revenue (Table 3.7).

Indeed the location of parent company operations and/or their major customers was found to have an important spatial significance in the initial geographical development and growth of such companies, as in the case of GSi, SG2 and CiSi. Thus SG2 opened branches where the Banque Societe Generale had major operations and similarly with CiSi and its parent the CEA.

Conclusions - Regional Development Implications: The regional consequences of these three processes of diversification, acquisition and externalisation are intimately bound up with trends towards concentration and internationalisation of services and are difficult to isolate in terms of their independent effects on the geographical development of services in Europe. Obviously all these issues require detailed research programmes to understand and gauge these processes fully on a regional basis. Nevertheless certain general points can be made here.

Increasing diversification in services leads to fewer but larger service market segments. This combined with acquisition activity is in turn associated with the fact that these enlarged service segments are being increasingly dominated by a few giant service corporations, and concern has already been noted relating to the effect of increasing concentration of ownership on the spatial concentration of strategic service functions and control on the less favoured regions of Europe.

The role of externalisation process is less obvious in a geographic sense and requires more detailed discussion, using some examples from the UK. A key to the growth of services within industry and the possibility of their externalisation or spin-off relates to their

TABLE 3.7 •

DANISH COMPUTER SERVICE COMPANIES HELD BY COMMON INTEREST GROUPS OR WITH A
MINIMUM 50% CAPTIVE REVENUE

COMPANY	OWNERSHIP	1985 REVENUE ($ million)
Kommunedata	Local Government	71.2
Data centralen	Independent non-profit making organisation, but depending on National Government demand	64.2
SpareKassernes Datacenter	Savings Bank	35.2
Landbrugets	Agricultural Co-operative	22.1

Source : Based on data from International Data Corporation (1986).

locational organisation within companies. It is expected that these service functions will be a mixture of planned and programmed activities (Thorngren 1970, 418-9) associated with a high level of <u>internal</u> contact activity. Examples of a planned function are corporate finance and central research and development, whilst an example of those programmed activities with a high internal contact content but more involved with routine, standardised processes in this context could be a computer service or data processing department. Studies in the UK by Parsons (1972), Goddard (1975), Crum and Gudgin (1977) and Howells (1984) of various functions within large corporations suggest that although more spatially dispersed, particularly on an urban, intra-regional level, than headquarters (orientation) activities these functions are nevertheless overall still overwhelmingly concentrated in the core South-East region of the UK and decline with increasing distance away from London.

Further growth and development of internal (to the firm) service functions would therefore appear to continue to entrench the concentration of service activity within European industry. This evolving pattern of internal service activity also has implications for the process of service externalisation and new service enterprise spin-offs. Although no research on this is available certain spatial development patterns can be suggested here. Given the location of internal service activities, noted above, most likely to generate service enterprise spin-offs, such as computer services, it is probable that these new service organisations will be highly concentrated in the main metropolitan areas of the Community. To a lesser extent, a number of key regional centres, such as Manchester in the UK or Lyons in France, with their concentration of regional headquarters, administrative activities and internal functions (such as data processing) may also gain to a more limited extent from such internal diversification processes.

Europe and the Geography of International Service Transactions

Introduction - Europe on Service Trade and Foreign Investment : This section will look at four aspects of international service transactions in relation to Europe. The first sub-section will outline the different elements in international service transactions and modes of operation before outlining the position of EC member states in services trade and foreign investment. The second will look at the growth and development of European multinational service companies, whilst the third will analyse the impact of foreign penetration on the EC domestic service market. The final sub-section will investigate the regional dimension of service trade, investment and multinational operations in Europe. Trade and foreign investment in information services will be considered in more detail in Chapter 4 (see also Howells 1987c).

It is useful to note the different ways in which services can be transacted on a national and international basis (Sapir and Lutz 1980; 1981). UNCTAD (1983, 6-7) have classified five different forms of service transactions. These are:

(1) Those services which are provided and consumed by residents of a country and do not enter into the international market place.

(2) Other services which are provided within national boundaries but to foreigners. These constitute international service transactions and include airport and seaport services rendered by domestic companies to foreign enterprises and the transport of foreigners by domestic carriers; tourism is another key example. Although these services are not exported the transactions have direct balance of payments effects.

(3) Services provided through direct export or import; for example, air freight and international re-insurance. This includes transactions between a domestic company and its overseas subsidiary.

(4) Services provided through contractual relationship:- associated with some form of royalty payments, fees or other remuneration associated with license or franchise relationships.

(5)   Services provided via foreign subsidiaries where certain services can only be supplied from within the country market and not through direct export, for example, hotels and banking.

Indeed Aaronovitch and Samson (1985, 152-3) go further and outline seven different modes of international service transaction in relation to insurance. They are :

(1)   Branch establishment.

(2)   Establishment of subsidiaries registered in the host country as legal entities. The effect of EC directives in bringing certain branches of insurance under supervision for the first time in some countries has a significant effect on establishing subsidiaries in the field of transport insurance in Germany for instance. Furthermore, non-EC insurance companies replaced branches of their main HOs by subsidiaries of HO set up in EC member countries.

(3)   Equity stakes in foreign countries whether simply as "trade investments" or to exercise some degree of influence or control. Such a stake may also be a prelude to a takeover bid.

(4)   Joint ventures such as Concorde-Minerve in Belgium or Norwich-Wintertur in the UK.

(5)   Collaborative/co-operative arrangements, such as AREA, UNISON or Campagnie I'internationale d'Assurances et de Reassurance.

(6)   Correspondents or agencies. All major companies have a substantial network or correspondents and agents, frequently brokers and broking groups.

(7)   Writing business in other countries from home base to the degree that is possible or allowed. It is important to note here that reinsurance is much less under supervisory control and is a means by which insurance can be written in this way.

Although these and other important distinctions can be made there are, however, essentially two main types of service transaction : those associated with the direct export of services and those associated with the sales of services via foreign affiliates. Any kind of real estimate of how important each are in terms of the international flow of services are extremely difficult (particularly in relation to transborder data flows; United Nations 1983) and even harder (or more spurious) for individual countries. However UNCTAD (1983, 45) have estimated that the international flow of services for Developed Market Economy Countries (DMECs) as a whole appears to be split roughly equally between export sales and sales of services via foreign affiliates or subsidiaries (indeed 86% of US service trade in 1976 was through foreign affiliates).

On a worldwide basis the value of service exports grew by 18.7% between 1970 and 1980, though this was a lower rate of growth than for merchandise exports which grew at 20.4%. Moreover by 1980 the value of service exports only represented 17.5% ($350 billion) of total exports (service and merchandise) valued at $2000 billion. Since 1980 service exports have grown more rapidly than merchandise exports, but even by 1983 service exports still only represented 18% of total world exports.

In 1980 three out of four of the world's largest service exporters were EC member states with the UK, France and West Germany ranked second, third and fourth behind the US. However, the UK had an export growth rate well below the world or DMECs average (Table 3.8), whilst France's growth rate was appreciably above the world average. Thus by 1983 the UK ($26.8 billion) was ranked fourth in terms of world service export trade behind the US ($39.4 billion), France ($38.5 billion) and West Germany ($27.7 billion). On a wider Community level eight EC countries were in the top 25 largest service exporters in the world in 1980 (UK, France, West Germany, Italy, Netherlands, Belgium, Spain and Greece; Table 3.9).

The relative strength of EC countries in terms of international service transactions would however appear to be less strong when considering sales of foreign service affiliates (Table 3.10; Column D)

TABLE 3.8

GROWTH IN TOTAL SERVICES TRADE, 1970-1980 : CREDIT/DEBIT

	Credit 1970-80 % Change	1980 Rank	Debit 1970-80 % Change	1980 Rank
World	611.0		623.1	
Developed market	574.1		567.0	
United States	520.5	1	420.3	1
United Kingdom	470.9	2	494.7	2
France	763.3	3	743.4	3
West Germany	597.0	4	579.7	4

Source : Howells and Green (1986a, 99; calculated from
data in : UNCTAD 1983).

TABLE 3.9

THE 25 LARGEST SERVICE EXPORTERS IN 1980

COUNTRY	VALUE OF SERVICES EXPORTS	VALUE OF FOREIGN INVESTMENT INCOME	VALUE OF MERCHANDISE EXPORTS	SERVICES BALANCE	SERVICES EXPORTS TO GDP RATIO (%)	SERVICES EXPORTS TO MERCHANDISE EXPORTS RATIO (%)
United States	34.9	70.2	224.3	6.0	1.4	15.6
United Kingdom	34.2	17.1	110.9	9.8	6.5	30.9
France	33.0	18.4	107.6	5.5	5.1	30.7
Germany	31.9	8.5	185.5	-17.9	3.9	17.2
Italy	22.4	5.3	76.8	6.2	5.7	0.2
Japan	18.9	7.2	126.8	-13.4	1.8	14.9
Netherlands	17.7	10.0	67.5	0.2	10.5	26.2
Belgium	14.5	17.6	55.2	0.5	12.1	26.3
Spain	11.7	0.2	20.5	6.3	5.6	56.9
Austria	10.8	2.5	17.2	5.1	14.0	62.6
Switzerland	8.4	NA	29.3	1.9	8.3	28.9
Sweden	7.5	0.8	30.7	0.5	6.0	24.3
Mexico	7.4	1.0	16.2	0.2	4.0	45.8
Norway	7.3	0.5	18.7	0.3	12.7	39.2
Canada	7.0	2.9	67.6	-2.5	2.7	10.3
Singapore	5.9	NA	18.2	3.1	54.1	32.7
Korea	4.5	0.3	17.2	0.6	7.7	26.1
Yugoslavia	4.5	0.2	9.0	-0.7	7.1	49.9
Greece	4.0	-	4.1	2.6	9.9	97.2
Saudi Arabia	3.7	NA	100.7	-8.0	3.2	3.7
Australia	3.5	0.7	21.7	-2.5	2.5	16.2
Israel	3.2	0.7	5.8	0.4	15.9	55.8
South Africa	3.0	0.4	25.5	-1.7	3.8	11.8
Finland	2.8	0.2	14.1	-	5.6	19.6
Egypt	2.3	0.3	3.9	0.1	9.8	60.2

Converted from SDRs and nominal values in national currencies at current exchange rates to US dollar; world is defined as IMF member countries reporting data for both 1970 and 1980.

Services exports exclude official transactions and investment earnings. Foreign investment earnings include private direct investment income and portfolio income, but exclude foreign official income.

Source : GATT (1985, 5).

TABLE 3.10

ESTIMATED WORLD TRADE AND FOREIGN DIRECT INVESTMENT IN SERVICES

COUNTRY	A EXPORTS OF SERVICES, 1980		B STOCK IN SERVICES, 1981		C OUTFLOWS, 1981-83 (YEARLY AVERAGES)		D SALES OF FOREIGN SERVICE AFFILIATES, 1982	
	$ billion	% of World total	$ billion	% of World total	$ billion	% of World total	$ billion	% of World total
USA	35	9.8	63	41.7	5	26.3	178	45.4
UK	34	9.5	13	8.6	3	15.8	32	8.2
FR Germany	32	8.9	11	7.3	1	5.3	27	6.9
Japan	19	5.3	18	11.9	5	26.3	44	11.2
Canada	7	1.9	6	4.0	-	-	14	3.6
Total above	127	-	111	-	14	-	295	-
Other developed market economies	165	46.1	35	23.2	5	26.3	86	21.9
Developing countries	66	18.4	5	3.3	-	-	12	3.1
World total	358	-	151	-	19	-	392	-

Source : based on Sauvant (1987, 284; see also Sauvant 1986a).

rather than direct export sales (Column A), where the US accounted for 45.5% of the world total of foreign service affiliates, followed by Japan with 11.2% (see Sauvant 1986a). Most interesting are the long-term implications of the growth in foreign direct investment by Japan (Column C).

European Multinational Service Companies and Foreign Investment : Nonetheless in world terms Europe is still relatively strong in terms of the number and size of its multinational service companies; stemming largely from its trading and colonial past. In particular the EC remains strong in financial services, especially banking and insurance. Thus seven out of the top twenty largest banks in the world in 1985 were from the European Community (Table 3.11). More generally a 149 (29.8%) of the top 500 largest banks in the world in 1985 were from the European Community with banks from Germany, Italy, France and the UK having a notable international presence (Table 3.12). Similarly in insurance German, French and UK companies have a very strong presence in the world insurance market, as seen in the number of offices they operate abroad (Table 3.13).

Other examples of EC multinational service companies are provided in Table 3.14. Retailing (C & A (NL), Storehouse Group (UK), Marks and Spencer (UK), Laura Ashley (UK)) is one service sector where European service companies are developing together with the more recent development of retail clothing franchises (Bennetton (I) and Pronuptia (Fr)). Another field is in holiday and business travel (Tjaereborg (Dk) and Club Mediteranee (Fr)) and related to this air transport (British Airways (UK), Lufthansa (FRG), Air France (Fr) and Aer Lingus (Ir)). In hotels, restaurants and catering Europe also has a significant and expanding presence (Grand Metropolitan (UK), IBIS Sphere (Fr), Vereinigung Interhotel (FRG), Steinberger (Fr), Accor (Fr) and Trusthouse Forte (UK)). Finally, engineering and building consultancy is also another sector where French, UK and German companies are also strong internationally (Sema 1984; Sema-Metra 1986).

TABLE 3.11

THE WORLD'S TOP TWENTY BANKS, 1985

BANK/NATIONALITY	NET ASSETS*	BANK/NATIONALITY	NET ASSETS*
1 Citicorp (US)	167	10 Credit Lyonnais (F)	111
2 Dai-Ichi Kangyo Bank (J)	157	11 Norinchukin Bank (J)	107
3 Fuji Bank (J)	142	12 National Westminster	
4 Sumitomo Bank (J)	135	Bank (UK)	104
5 Mitsubishi Bank (J)	132	13 Ind. Bank of Japan (J)	103
6 Banque National de Paris (F)	123	14 Societe Generale (F)	98
7 Sanwa Bank (J)	123	15 Deutsche Bank (D)	95
8 Credit Agricole (F)	123	16 Barclays Bank (UK)	94
9 Bank America (US)	115	17 Tokai Bank (J)	90
		18 Mitsui Bank (J)	89
		19 Chase Manhatten (US)	85
		20 Midland Bank (UK)	84

* Less contra accounts in US $'000 million.

Source : Compiled from data in The Banker, June 1986, 69-203.

TABLE 3.12

NUMBER OF BANKS BY SELECTED COUNTRIES IN THE LARGEST[*]
500 BANKS, 1985

COUNTRY	NUMBER OF BANKS
United States	110
Japan	77
EC :	
West Germany	42
Italy	27
France	18
United Kingdom	16
Spain	12
Belgium	8
Denmark	7
Luxembourg	6
Netherlands	4
Portugal	4
Greece	3
Ireland	2
TOTAL for European Community	149

*   Based on assets.

Source : Compiled from data in The Banker, July 1986,
69-203.

TABLE 3.13

INTERNATIONAL OPERATIONS OF UK AND OTHER INSURANCE COMPANIES 1982 : TOP FIVE COUNTRIES

	NO. OF COUNTRIES WHERE OPERATING	NO. OF OFFICES ABROAD
UK	47	675
US	47	609
France	33	185
Switzerland	27	144
West Germany	18	119

Source : Sigma.

TABLE 3.14 : EXAMPLES OF MAJOR EUROPEAN MULTINATIONAL SERVICE COMPANIES

COMPANY NAME	NATIONAL	SERVICE ACTIVITY	RANGE OF ESTABLISHMENTS LOCATION	OTHER COMMENTS eg. ACQUISITIONS
				Acquired :
Allianz	FR Germany	Insurance	--	Cornhill Ins., UK and Riunione Adriatica di' Sicurta, Italy
Bennetton	Italy	Clothing retail franchise	Throughout EC and US (71 principle agents)	--
Cap-Gemini-Sogeti	France	Computer services	Throughout EC and US eg. US: CAP Gemini Dasd Inc.; Cap Gemini Services Inc.; Cap Gemini Software Products Inc.; ICOMX; CGA Computers	Acquisitions in UK, FR Germany, Italy and Netherlands
Tjaereborg	Denmark	Holiday travel	Throughout EC	--
C & A Brenninkmeyer	Netherlands	Retailing & related manufacture	Throughout EC	Operates 20 manufacturing plants via its Canada Int. subsidiary
Laura Ashley	UK	Retailing & related manufacture	225 shops worldwide in Europe, N. America and Australia	--
Club Mediteranee	France	Holiday travel and related interests	Throughout EC and world	--
Saatchi and Saatchi	UK	Advertising and business services	Throughout EC and world	Largest advertising agency in the world
Lambert SA		financial services		Henry Ansbacher UK merchant bank
Aer Lingus	Ireland	Air transport, hotels & related interests		Owns Dumfrey Hotels, 41 hotels worldwide
Accor	France	Hotels	500 hotels in 70 countries	Created via merger of Novotel and JBI £1.6 billion 1985 turnover

EC multinational service corporations, due to internal barriers to trade and their former colonial status, have generally been much slower to expand within the European Community than outside it, particularly in relation to North America. This is gradually changing in certain sectors such as retailing where specific pan-European operations are being established by UK and other companies. Recent acquisition activity by European companies, such as Deutsche Bank's takeover of Banca d'America d'Italia in banking and Allianz's recent moves in insurance are also changing this pattern. Indeed in certain sectors such as insurance, joint European consortia (eg. AREA - la Royal Belge, Eagle Star Allianz and AMEU-Utrecht) are being developed in order to meet the growing insurance requirements of European multinational companies.

However, of concern here is that are very much fewer truly international European companies operating in the newly emerging, novel and innovative service sectors. This includes the new fields of computer, business and information related service sectors, associated with management consultancy, advertising, computer software, accountancy and electronic information services. There are a few exceptions to this, for example, French computer software companies, as noted earlier, which are building up a strong international presence, whilst in accountancy UK-based partnerships still hold a large share (an estimated 10%) of the world market. However, overall it is in these sectors where US firms are gaining an increasing foothold both on a global scale and within the Community.

Foreign Penetration and Investment in the European Service Market : As noted above the increasing domination of US companies in new and emerging sectors of business and information services is seen in their increasing foreign investment and acquisition in the EC market for these types of services (Table 3.15). Thus it has been estimated that over half of the on-line financial information services market in Europe is controlled by non-European, mainly US suppliers (Chapter 4). Similarly, US companies dwarf their European counter-parts in terms of (1984)

TABLE 3.15 : EXAMPLES OF NON-EC SERVICE COMPANIES OPERATING IN EUROPE

NAME	NATIONALITY	SECTOR	EXAMPLES OF SUBSIDIARY COMPANIES IN EEC
Dun & Bradstreet Corporation	US	Information services	Dun & Bradstreet Computer Service (France) SA (Fr) Dun & Bradstreet Lusitana Ltsla (P) Dun & Bradstreet SpA (I) Neodata Services Gmbh (FRG) AC Nielsen Mellas Ltd (Gr) Datastream Plc (UK)
IMS International Inc.	US	Information services	IMS Italiana SpA (I) Pharmex Computer Gmbh (FRG) Informations Medicales et Statistiques S.A.R.L. (Fr) Mercados y Analisis SA (S) Investigacoes Mercadologicus e Sociologicus Ltd a (P)
TMT Limited	Australian	Transport	TNT Express Frachtdienste Gmbh (FRG) TNT IPEC Denmark ApS (DK) TNT Skypak (Mellas) Ltd (Gr) Thomas Nationwide Transport France S.A.R.L. (Fr) TNT-IPEC Insurance BV (NL)
Fuji Bank	Japanese	Banking/ financial services	Mellas Factoring Espanola SA (Sp) Cofracredit (Fr) Nordisk Factoring A/S (DK) Mellas Factoring Bank AG (FRG) Eurocapital SA (Lux)
Citicorp	US	Banking/ financial services	Banco Centro Sud SpA (I) Banca de Biase e.l. SpA (I) Vickers da Costa Ltd (UK) Diners Club of Greece SA (Gr) Famicredit (Fr) International France Associates BU (NL)
News Corporation Limited	Australian	Publishing/	Falconworld plc (UK) incl. AB Computerlink Ltd (UK) TCF Holdings Inc. (USA) incl. Produzione Artistiche Internationali (I) Foxfilmes Ltda (P) Hispano Fox Film SAE (Sp) Newscorp Investment Ltd (UK) incl. News International plc (UK) Media International SA (Bel) (with Newscorp Services BV (NL) Satellite Television plc (UK) News Satellite Television Ltd (UK) The Times Ltd (UK)

worldwide computer service sales (IBM $3.4 billion, ADP $0.96 billion, Control Data $0.93 billion, TRW $0.92 billion and EDS (GM) $0.77 billion) compared with the five largest European companies (Cap-Gemini-Sogeti $207 million, CiSi $173 million, Finsiel $155 million, SG2 $151 million, Scicon International $138 million) with indeed IBM dominating the European computer software market, in terms of both direct and third party sales (with computer software sales of $4.5 billion by 1985).

In other sectors however, foreign penetration of domestic service markets have been highly restricted or indeed totally excluded. The case of German insurance was one obvious example of these restrictions, where German regulations forbade foreign insurers to do German business unless they were physically established there. However the European Court of Justice has recently ruled that Germany (and France, Ireland and Denmark) have broken the Twenty of Rome's rules in free trade in services by insisting that foreign insurers had to be registered locally to write policies for local clients. This verdict on the requirement of permanent establishment has far-reaching consequences in terms of opening up the market for insurance and other services market in the Community. Other highly restricted service sectors cover professional services where legal, health and education professions have been virtually excluded from non-domestic participants. However, recent initiatives by the Community has helped open up certain professions, such as architecture, to individuals of other member states countries to operate more easily in their markets.

It is important to distinguish between foreign _entry_ and foreign _penetration_ in regional development terms. As in the case of the City of London the _entry_ of foreign banks, in terms of their setting up offices or joint ventures and consortium, in the UK indicates the important role that London now plays in the international financial and banking markets together with New York and Tokyo (Table 3.16). London has therefore become a key location for the operation of foreign banks worldwide and foreign entry into the UK is part of this positive process, providing additional jobs (now over 50,000 people). Similarly

TABLE 3.16

GLOBAL FINANCIAL CENTRES : LONDON - NEW YORK - TOKYO AXIS

	LONDON	NEW YORK	TOKYO
Foreign banks	399	254	76
Foreign SE* members	22#	33	6
Foreign dealers in primary government bond market	12#	2	29
Share of International banking market (1985) %	24.9	15	9.1
Share of foreign exchange turnover (1984) %	32.6	23.3	5.3
Stock market turnover £bn (1985)	52.8	671.3	271.5
Stock market capitalisation £bn (end 1985)	244.7	1,302.2	648.7

* Stock Exchange

including UK firms partly or wholly foreign-owned.

Source : Financial Times, 7 April 1986.

Lloyd's in London is the single most important 'market' for general insurance in the world (Aaronovitch and Samson 1985, 93), with about two-thirds of Lloyd's business being in reinsurance, although its pre-eminence is being threatened by the growth of alternative insurance centres now being created in the United States.

However, even here the positive aspects of employment growth, for example, have to be tempered by the fact that UK banks for example have since 1980 become less important in the London international lending market (Howells and Green 1986a, 113). Thus, Japanese banks account for over a third of all London's overseas lending in non-sterling currencies. Indeed the Bank of Tokyo, Fuji, the Industrial Bank of Japan and Dai-ichi Kangyo now hold nearly a quarter of all banking assets in the UK which is only just behind the UK clearing banks.

Indeed in relation to Japanese investment what has been overlooked is that the majority of their direct inward investment into Europe has not been into manufacturing industry but rather into services (Table 3.17). Thus investment in manufacturing in Europe by Japanese companies is less than 20% of total Japanese direct investment, compared with nearly 70% for services. Indeed total manufacturing (18.9%) is smaller than either finance and insurance (33.6%) or commerce sectors on their own (22.6%).

Foreign penetration of national and regional domestic service markets, by contrast, may have very much more profound negative impacts on the local and regional economy in terms of the development of 'branch plant' service economies with the consequent loss of higher order, more dynamic and innovative service functions. There are some positive aspects of foreign penetration. It may improve the competitive environment by breaking up an oligopolistic market structure at a local or regional level and it may provide the service more cheaply and efficiently (Aronson and Cowhey 1984, 18) than a smaller indigenous competitor. However, partly because of this, the overall impact of foreign penetration of service markets is likely to reduce employment prospects as foreign companies (by definition more dynamic because they have passed a key 'organisational threshold' in terms of their ability to locate abroad; Taylor 1975, 314-7) provide services more efficiently and push indigenous companies out.

66

TABLE 3.17

JAPAN'S DIRECT OVERSEAS INVESTMENT IN EUROPE

	SECTOR	$ million*	% of Final Total
1	Total Manufacturing	2,088	18.9
2	Commerce	2,486	22.6
3	Finance & Insurance	3,695	33.6
4	General Services	252	2.3
5	Transportation	29	0.3
6	Real Estate	57	0.5
7	Others	1,027	9.3
8	Total Services (2-7)	7,546	68.6
9	Combined Total (1+8)	9,634	87.6
10	Final Total (including investment in Agriculture, Mining, etc.)	11,002	

* As at March 31, 1986.

Source : Compiled from data from the Ministry of Finance, Japan.

International Service Transactions and Regional Development —
Conclusions: In regional development terms it is important to take the
long term net effects of international service transactions. It is
suggested here that the impact of services and establishment of foreign
service establishments will be highest in the core regions of a country.
Thus Dunning and Norman (1983) found that the location of foreign
service establishments in Belgium, France and the UK were overwhelmingly
concentrated in the national capital cities of their respective
countries (Table 3.18). Research by Daniels (1986b) on foreign banking
activity in the UK and USA also confirms this finding and reveals a
marked concentration of foreign banking operations in London and, to a
lesser extent, New York. Thus of all the foreign banks with operations
in London only 18% have offices elsewhere in the UK, although a more
decentralised pattern might emerge if foreign banks continue to increase
their representation in the UK (Daniels 1986b, 276-7 and 285). As such
foreign penetration of service markets in peripheral, less favoured
regions of national territories is likely to be very much lower as these
are generally smaller, less lucrative and more inaccessible markets.
However, when considering the net effects (taking into account the
effects of service exports and foreign investment) it is presented here
that core regions will be net beneficiaries, ie. positive invisible
service transactions balances, whilst peripheral, less favoured regions
will generally have negative balances. Both inflows and outflows of
service transactions (trade and investment will therefore will be at a
much higher level in core regions but have a positive net balance,
whilst in peripheral regions inflows and outflows of service
transactions are much lower and tend to have a negative net balance.
Transaction flows can obviously not only refer here to trade and
investment by the region on an inter- national basis, between the region
and foreign countries, but also on an inter-regional level, between the
region and other regions in the country.

Detailed research needs to be undertaken into this whole issue, but
above all it needs to be recognised that firstly the level of
interregional trade in services, and therefore its consequent impact on
the regional economy, is much greater than is generally assumed. For

TABLE 3.18

DISTRIBUTION IN TERMS OF NUMBERS OF OFFICES OF US-BASED BUSINESS SERVICES

BUSINESS SERVICE	BELGIUM		FRANCE			UK[a]		
	Brussels[b]	rest	Paris central[c]	Paris non-central	rest	London central[d]	London non-central	rest
Management consultancy and executive search	39	2	19	9	–	40	9	3
Advertising	13	1	13	6	–	15	4	–
Accountancy	9	2	2	5	–	12	–	–
Insurance	20	5	6	6	–	16	4	1
Banking	18	3	32	7	–	73	1	2
Engineering design	14	3	9	7	1	8	8	2
Legal practice	10	–	–	–	–	24	3	–
Others	28	2	4	–	8	25	5	6
Subtotals	151	18	85	40	9	213	34	14
Grand Totals	169		134			261		

a These figures may understate the number of business service activities in the UK since they relate solely to US firms in the UK that are sustaining or active members of the US Chamber of Commerce.

b A 'central/non-central' division was not possible for Brussels.

c 1st-9th Arondissements.

d All EC and WC areas, plus W1.

Source : Dunning and Norman (1983, 198) compiled from [US Chambers of Commerce in Belgium, France and the UK (UK figures are for 1976; Belgium and France figures are for 1977)]

example, Polese (1982, 158; see also Polese 1981) found that over half of regional service demand in a rural area of Quebec was satisfied by imports. Secondly, related to this, is the significance of export-led growth in regional development. Thus Beyers and Alvine (1985; see also Beyers et al. 1985; 1986) in a survey of service firms in the central Puget Sound area of Seattle, USA found that 1105 firms out of 2000 'exported' more than 10% of their output outside the Puget Sound area.

Clearly therefore the significance of service transactions on regional economic development should not be underestimated (Dinteren, 1987). Strategies therefore designed to improve the net service transaction flow of a region via, on the one hand, import substitution of services and, on the other, through the encouragement and development of export-oriented services can have a significant impact on the economic growth and development of a region.

Deregulation, Privatisation and Barriers to Trade in Services

Although in the past the service sector in Europe has been highly protected and regulated, with often a high degree of state involvement in terms of control of nationalised companies operating in service sectors, this pattern, as noted earlier, appears to be changing. The lead in this field was taken by the UK with the arrival of a Conservative government in 1979, committed to a laissez-faire economic approach with as little state intervention and control as possible. A wide ranging programme of privatisation was implemented which included selling off stakes in a whole series of corporations which operated in the service sector (Table 3.19). More recently other countries have also moved towards privatising or de-nationalising key public sector enterprises with the governments of Belgium, the Netherlands and Italy all reducing their stakes in their national airlines. In France a programme has been produced for the denationalisation of a number of public enterprises covering major banks, insurance and financial holding companies (Table 3.20). The government has also privatised a number of other service corporations including : Havas, the advertising and tourist group, for FFr 4 billion ($670 million) and TF-1, the television

TABLE 3.19

PRIVATISATION OF UK SERVICE CORPORATIONS

NAME OF CORPORATION	SERVICE SECTOR INVOLVED	PROGRESS IN PRIVATISATION	PROCEEDS £m
British Telecom	Telecommunications	Completed	3,900
National Freight Consortium	Road transport/ distribution	Completed	5
BAA plc (formerly British Airports Authority (BAA))	Major UK Airports	Completed	1,200
ssociated British Ports	Major UK Seaports	Completed	94
British Airways	Air transport	Completed	900
Sealink (formerly part of British Rail)	Passenger ferries	Completed	66
British Rail Hotels (formerly part of British Rail)	Hotels	Completed	45
Thomas Cook	Travel agents/ Tour Operators	Completed (sold to Midland Bank)	
National Bus	Public transport	In progress (local subsidiaries being sold seperately)	200[*]
Cable and Wireless plc	Telecommunications	Completed	1,015
Trustee Savings Bank[**]	Bank	Completed	

[*] Estimated

[**] Based on a House of Lords decision that ownership rested with the Crown (ie. the State/National Government). It was decided that no proceeds would be obtained.

Source: updated and revised from Howells and Green (1988).

TABLE 3.20

PLANNED PRIVATISATION OF FRENCH SERVICE CORPORATIONS

BANKS	PROFITS (1984)	(FFrbn) NET ASSETS (1984)	MARKET VALUE ESTIMATES (1985)
BNP	+ 1.7	948.5	21.3
Societe Generale[+]	+ 1.2	835.8	15 (22.4[*])
Credit Lyonnais	+ 1	868	13
CCF	+ 0.2	160	2.8
Credit du Nord	+ 0.3	0.9	0.6

INSURANCE COMPANIES	PROFITS	NET ASSETS	MARKET VALUE
UAP	+ 1.3	28	8.8
GAN	+ 0.6	15.8	4.1
AGF	+ 0.95	18	5.7

FINANCIAL HOLDING COMPANIES	PROFITS	NET ASSETS	MARKET VALUE
Paribas[+]	+ 1.1	539	11.7
Group Suez[+]	+ 0.7	263	6.9

+ Privatisation completed
* 1987 privatisation valuation

Source : Data from French Industry Ministry and current information.

channel for FFr 3.465 billion ($570 billion). More recently the French government is also expanding Air France asset base via a 15% expansion of its share which will be sold to the public. Similarly in Germany the government's privatisation programme has been partially revived, although selling off part of the government's holding in Lufthansa, the national airline, has been shelved. Other more ad hoc sell-offs of public enterprises have also been undertaken by national governments, for example, in Spain (Table 3.21).

In relation to deregulation and reducing barriers to trade in services, progress has been much slower. Again the UK has moved most rapidly on the deregulation of key service sectors such as telecommunications, financial markets and public transport in conjunction with its privatisation plans. However, in other European countries there have also been movements towards reducing the regulatory environments of telecommunications (Table 3.22) and financial markets and countries have been involved in bilateral negotiations in opening up competition on air routes. Indeed, in a fundamental sense, deregulation is increasingly being used as a competitive tool between countries to attract locationally footloose financial and business investment by large multinationals (Chapters 4 and 5). Thus as Dyson (1986, 28) has recently noted "The threat of location of investment by multinationals in more deregulated environments promotes a process of competitive deregulation. Deregulation offers the prize of investment and jobs in financial services, telecommunications and broadcasting and increased tax revenue from these sources".

The trend towards the removal of barriers to trade and investment in services has been associated with wider negotiations within the European Community on the General Agreement on Tariffs and Trade (GATT) round of negotiations and opening up the internal market in services within the Community (Krommenacker 1987). Evidence from work on this issue (Peat Marwick, Mitchell & Co. 1986) suggests that barriers to trade are wide-ranging although in some sectors (for example, advertising) there are fewer restrictions than others, such as

COUNTRY	DEREGULATION	PRIVATISATION	FURTHER DETAILS
Spain	--	*	Partial privatisation of the PTT together with a number of public service corporations, eg. Marsans (travel agents) Entursa (hotels).
Italy	*	--	Deregulation of banking/finance, however, re-regulation of many more sectors, eg. TV.
Portugal	--	*	Partial privatisation in banking/insurance.
Greece	--	--	Strong moves since 1981 towards nationalisation.
Netherlands	*	**	Privatisation limited as yet to a few peripheral activities, eg. weights/measures, port pilot services. However major deregulation of the PTT planned for 1989; together with air travel deregulation.
Belgium	*	*	Privatisation – RMT (sea transport); PTT-proposed. Deregulation – trading laws, international HQ operations.
Denmark	--	--	Use of private services highly variable between commune – no real change.
Ireland	**	--	Deregulation of financial/banking services.
France	**	**/***	Privatisation in progress in banking insurance and media (Table 3.20).
FR Germany	--	*	Privatisation locally centred : mainly public transport, refuge collection.
UK	***	***	Major privatisation and deregulation programme in progress covering service activities (see Table 3.19).

Notes : -- No clear initiatives on privatisation/deregulation.
     * A very limited initiatives on privatisation/deregulation.
    ** A limited initiative on privatisation/deregulation.
   *** Major initiatives on privatisation/deregulation.

TABLE 3.22

TERMINATION SERVICE STRUCTURES IN THE EUROPEAN COMMUNITIES

	Basic Service Network				Use of Leased Circuits			
					Domestic		International	
	Local	Long Distance	International	Mobile	Shared Use/ Resale	Public Network Interchange	Shared Use/ Resale	Public Network Interconnection
Belgium	GM(PC)	GM(PC)	GM(PC)	GM(PC)	GM(PC)	N	N(1)	N(1)
Denmark	OM(2)	OM	GM	OM	N	N	N(1)	N(1)
France	GM	GM	GM	GM(3)	N(4)	N(4)	N(4)	N(4)
FR Germany	GM	GM	GM	GM	Y(5a)	Y(5b)	Y(5a)	Y(5c)
Greece	GM(PC)	GM(PC)	GM(PC)	PL	N	N	N(1)	N(1)
Ireland	GR(PC)	GM(PC)	GM(PC)	GM(PC)	N(1)	N(1)	N(1)	N(1)
Italy	GM(PC)	GM(PC)	GM(PC)	GM(PC)	N(6)	N(6)	N	N
Luxembourg	GM	GM	GM	GM	N	-	N(1)	N(1)
Netherlands	GM(7)	GM(7)	GM(7)	GM(7)	N(8)	N(8)	N(1)	N(1)
Portugal	GM(PC)	GM(PC)	GM(PC+OM)(2)	-	N(9)	N(9)	N(1)	N(1)
Spain	OM(10)	OM(10)	OM(10)	OM(10)	N	N	N(1)	N(1)
UK	RC(LIM)	RC(LIM)	RC(LIM)	RC(LIM)	Y(11)	Y(11)	Y(12)	Y(12)

Key:

PL	Partly liberalised	RC(LIB)	Regulated competition with liberalised entry
GM	Government monopoly (government agency)	FC(LIB)	Free competition with liberalised entry
GM(PC)	Government monopoly (public corporation)	Y	Generally permitted
OM	Monopoly of other types (private entity, etc.)	N	Generally prohibited
RC(LIM)	Regulated competition with limited entry		

Notes
(1) Subject to exceptions
(2) Telecommunications service providers exist in to PTT, on a monopolistic basis (concessionary basis, regional monopoly, etc).
(3) Licensing of additional providers to be announced.
(4) Steps regarding licensing of private providers of value-added services announced.
(5a) Shared use permitted, resale prohibited.
(5b) Voice-band circuits: as far as technically possible, but at one end only (TKO, July 1986).
(5c) International fixed connections without restrictions; 'flat-rate' circuits with restrictions.
(6) New legislation of VANS is being discussed in parliament.

(7) PTT to be converted to limited liability company in 1989.
(8) Usage for VANS to be liberalised.
(9) Currently under consideration in commissions.
(10) Telex, telegram, public facsimile (Burofax), etc are provided by the PTT.
(11) Pure resale prohibited until at least 1989.
(12) As (11) subject to additional restrictions.

Source: Compiled from Senior Officials Group on Telecommunications (SOG-T returns in Commission of the European Communities, 1987, Figure 7).

telecommunications. Barriers also take different forms. Thus, for example in insurance, although there are relatively few barriers to trade there are generally much greater obstacles to foreign investment and establishment via licensing procedures. However, again there are substantial variations between countries. Some countries such as the UK and the Netherlands are more 'open' in terms of their insurance market than others such as Germany, Ireland, France or Denmark (Aaronovitch and Sawyer 1985, 167-8). Similarly in banking a country, such as the Netherlands (where for example a third of the Dutch banking market is held by foreign companies), has a very open banking system, compared with other EC countries where foreign banking is greatly restricted.

All these changes in privatisation, deregulation and removal of barriers to trade and investment have important implications for locational change in services and for regional development policy. Thus privatisation and deregulation of telecommunications and transport services are likely to have important implications for less favoured regions in Europe associated with, for example, the loss or downgrading of the role of cross-subsidisation in service provision. Similarly, network providers may no longer be required to maintain services to an equal standard across the national territory, above a certain basic minimum requirement. Thus in telecommunications the provision of new, more sophisticated services will be restricted to core areas of countries. An example of this has been the Highlands and Islands Development Board disappointment in the lack of development in terms of the provision of new telecommunication services to that region.

Similarly, removal of barriers to service trade both internally within the Community and externally with its major trading partners will have a profound effect on regional development in Europe. Indeed there is a view that free trade and foreign investment in services will not benefit one whole country at the expense of another, but it is more likely to give obvious advantages to one group or region within a country whilst conferring a more generalised disadvantage to other groups or regions. As such the economic consequences of liberalising trade in services is more likely to be felt at a _regional_ rather than a national level with the Community.

More specifically in relation to barriers to service trade, bilateral or multi-lateral trade agreements can take the form of eliminating or reducing both tariff and non-tariff barriers to free trade and foreign investment. Agreements could include a general reduction in-barriers across the whole range of services or could be selective. In either case the national and regional effects are likely to differ across the Community. For example, an 'across-the-board' reduction to barriers in trade for services may have limited effects in the less favoured regions of Greece but have widespread implications for the core regions of Germany. A reduction or abolition of controls on air fares, through its impacts on tourism may however have a relatively greater effect on peripheral localities in Greece.

The effects are also likely to be different dependent on the timescale over which any adjustment takes place. If barriers are suddenly removed then those areas with an existing advantage are likely to reap the full benefit of removal. If on the other hand adjustment takes place over a long period of time the existing advantage will be more muted and indigenous industry may be able to accommodate the change more readily. At the national level a country may therefore achieve advantages in the increased availability and efficiency of services without the destruction of existing or embryonic indigenous service provision.

At the regional scale the effects of changes in barriers to trade and foreign investment in services may be both direct and indirect (Table 3.23). For example, in the UK Northern Region may directly receive more efficient business services as a result of trade liberalisation which may have a positive effect on industry located there. On the other hand, this same competition may indirectly reduce the foreign earnings of the business service sector in the Northern region and the UK as a whole, worsening the Northern region's and the UK's balance of payments leading to deflation which in turn results in

TABLE 3.23

IMPACT OF TRADE AND FOREIGN INVESTMENT LIBERALISATION IN
SERVICES ON REGIONAL ECONOMIES

a)   Service Suppliers

Positive   (1) increase the export potential of existing service supply
               industry (eg. tourism, shipping services, etc.).

           (2) increase inward investment to the region by service
               providers.

Negative   (1) undermine directly the service supply industry as a
               result of competition from external suppliers.

           (2) lead to takeovers of existing syppliers by external
               suppliers, thereby reducing indigenous capacity.

b)   Service Consumers

Positive   (1) improve the competitive strength of local economic
               activities (primary, secondary and tertiary sectors)
               though cheaper, more economically efficient,
               qualitatively enhanced or completely new services.

Negative   (1) paradoxically free trade and competition may reduce
               choice and increase the prices of services in the long-
               term. This might arise via the eventual domination of
               key service markets by a few giant (oligopolistic)
               multinational corporations developing out of a process
               of concentration in service activities related to
               merger/acquisition activity and the disappearance of
               smaller uncompetitive intra-national service companies.
               Few national governments consider regional monopolies or
               the regional consequences of mergers.

rising unemployment. Moreover, by removing restrictions on foreign investment in service activities peripheral regions may find their most successful and innovative service companies being taken over by large multilocational service companies centred in the core regions of Europe or from the US and Japan. This could lead to a loss of indigenous potential and to the formation of 'branch plant' service economies in the regions associated with external control and the provision of routine, low cost service activities. As such, service establishments in these more peripheral areas, coming under the control of large corporations centred outside the region, will be associated with the delivery of less sophisticated, routine services whilst new high-order, generative functions will be transferred to the head offices located in the core regions of national territories. Not only will less favoured regions lose the ownership, control and strategic-making functions of such service industries, but in turn these externally controlled service establishments will reduce their own resource requirements (input linkages) from local or regional sources and instead satisfy them from the head office location located outside the region (Leigh and North 1978; Marshall 1982).

As such, consideration needs to be given to the balance of positive and negative effects on the existing market for service industries in the less favoured regions of the Community. Increased possiblities for trade can of course lead to increased imports as well as to new opportunities for export. We can assume that the consumers of market services will benefit from increased competition through lower prices or enhanced existing or entirely new services. Some assessment needs to be made of the balance of positive and negative effects which take into account both the supply industry (as a source of jobs and wealth creation) as well as service consumers. The types of positive and negative impact are summarised in Table 3.23.

Finally it should be stressed again that increased trade flows and direct investment will occur not only on an international basis within the Community but also on an international level between the Community and non-EC countries, in particular the US and Japan. If this latter international trade grows more rapidly and yields a net deficit to EC countries, national and regional economies within the Community will obviously suffer. In particular there is the problem of the encroachment of non-EC companies on domestic EC service markets via direct foreign investment. An example of this is in the UK where Japanese and US banks and financial institutions have expanded rapidly, associated in part with the integration and liberalisation of the London financial markets. EC banks and financial houses on the other hand, with the exception of some French and German banks, have expanded much more slowly in the UK.

New and Small Service Firms and the Informal Economy in Europe

New and Small Service Firms in Europe : Very little research has been undertaken on new or small service firms in the European Community. This is surprising given the fact that, although most commentators agree that new firm formation in manufacturing industry will have little employment or production impact over the short and medium run, in private sector services this is much less true. Thus Keeble and Wever (1986, 25-6) argue that new firms in many of the growing private sector service industries will probably have quite considerable job creation and output significance even over the short run. This is suggested by the low average firm size and entry barriers in many consumer and producer service activities, which in turn explains the high rate of new service firm formation. It is also confirmed by the recent growth in the net surplus (births minus deaths) of new service firms in countries such as the UK (Howells and Green 1986a, 71-75; Keeble and Weaver 1986, 26).

However, it has been noted in the UK that although the service sector was a expanding source of new and small firms this trend is taking place within an increasing level of industrial concentration within service sectors (Howells and Green 1986a, 71). Increasing concentration levels in services has also been confirmed in Germany with a gradual decline in small firm share of service employment (Hull 1983, 157-9). The importance of new and small firms in services therefore appears to continue despite, or paradoxically because of, increasing concentration ratios within such industrial sectors, as individual sectors become dominated by a number of large companies. This in turn would confirm the dual-economy model which suggests the development in specific industrial sectors of a small number of giant business organisations being supported by a range of new and small companies (Taylor and Thrift 1983). Support for this hypothesis in the UK comes from an employment study of small firms (Rajan and Pearson 1986, 135) which suggests that business growth in small firms has emanated from three sources : discovery of new market niches; subcontracting by larger firms; and the displacement of business outside the small firms sector.

Data from the European Community would appear to also substantiate this dual hypothesis view of the service sector. Korte (1986, 40) analysing the recent distribution of employment by size of enterprise in the tertiary sector shows that in comparison to the distribution in manufacturing, employment is much more polarised (Figure 3.1). In all Member States for which data were available, establishments with 10-49 and 1000 employees and over have the highest employment shares. By contrast employment in establishments within the four groups covering the 50-999 employee size range hardly exceeded 10% of the total in each case. Thus employment in the '10-49' and the '1000 and over' groups accounts in both cases for over 20% in all Member States, even exceeding 30% in most of them.

In short therefore although new and small firms represent a significant share of service employment in Europe this would appear to be within a wider context of increasing industrial concentration. This latter fact is indicated by studies for example from the UK (Howells and Green 1986a), Germany (Hull 1983; Bade 1986) and Greece (Dokopoulou 1986). Hull (1983, 157-9), for example, found that there was a gradual decline in the small firm share of service employment in Germany between 1950 and 1970, whilst Bade (1986, 265) analysing sales by size class between 1970 and 1982 found no shift towards smaller firms due to the

FIGURE 3.1

DISTRIBUTION OF EMPLOYMENT IN ENTERPRISES AND ESTABLISHMENTS
BY THEIR SIZE AT EUROPEAN LEVEL 1978*

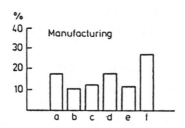

a)   10 -   49 employees   d)   200 -   499 employees

b)   50 -   99 employees   e)   500 -   1000 employees

c) 1000 - 199 employees   f)   1000 and over employees

*   The EEC-average for the service sector foes not include Greece, the
    U.K., Ireland and Luxembourg. The EEC-average for manufacturing
    excludes Ireland, Luxembourg and Greece.

Source: Korte (1986, 40) using data from EUROSTAT, Labour Cost Survey
          1978. rapid expansion of large firms within the service sector.
Similarly Dokopoulou (1986, 300-1) although noting that small
establishment growth was most rapid in service sectors, overall small

82

establishments employment growth rates between 1960 and 1978 were lower than for all establishments.

More qualitative aspects should also be considered in relation to the significance of new and small firms in the European service economy. Thus high birth rates and low entry barriers in retailing, catering and wholesaling (Howells and Green 1986a 75) appear to be in those sectors which have less long-term significance to their local and regional economies because of their largely non-basic (local servicing and low trade/export potential) character and consequent high dependency on the growth prospects of their local area and low level of technological sophistication. Support for this view appears to come from Korte (1986, 42) who found that the employment growth of small and medium sized enterprises (SMEs) has occurred mostly in service industries such as wholesaling and retailing, whilst most high productivity services, like banking, exhibited increasing employment in large enterprises.

In the context of the spatial distribution of new and small firms in the most innovative, technologically advanced services evidence from France (Aydalot 1986) and the UK (Mason 1985; Howells and Green 1986a) suggests that such firms are highly concentrated in the core regions and attractive areas of the national territories. Thus in France firm creation in most advanced service sectors was centred on Paris, with 45% of the new firm formation growth rate between 1974-1983, with southern sunbelt regions being also highly placed (Aydalot 1986, 114). Evidence from the UK also suggests that there are considerable core-periphery variations in the location of dynamic, innovative new and small firms. Mason (1985, 1508) in a study of successful small firms in the UK concluded that the analysis indicated that the peripheral regions in the UK are handicapped by their acute shortage of Small and Medium Sized Enterprises (SMEs) that can be defined as successful either on technological or on financial criteria and, by implication, growing rapidly. Similarly Howells and Green (1986a, 155) concluded that in terms of new service firm formation rates and more particularly the survival, growth and innovativeness of small service firms the peripheral regions in the UK have much poorer performance records.

Informal Economy in Europe : Much of the report here necessarily uses government data in relation to the size and growth of the service sector. It should be recognised, however, that a relatively small but significant part of the economy goes unrecorded and this is particularly true of the construction and service industries. The 'informal' or 'black' economy has been variously defined depending on which aspect of the phenomenon is most important, i.e. whether it is seen mainly as an officially unrecorded activity, transaction or income (Lambooy and Renoy 1985, 3). In relation to it as an activity it can be defined as the production of goods and services for gain but concealed from official notice in order to avoid payment of tax and other dues to the state (Mason and Harrison 1985 19).

Estimates of the size of the black economy obviously vary according to definition and measurement. The size of the black economy in the UK, for example, estimated from the national accounts perspective, was between 2% and 7.5% of GDP (according to the definition adopted) whilst on a fiscal basis it was estimated between 2% and 3% of GDP (Smith 1981). Other work comparing the UK in an international context has put the size (this time based on GNP) of the black economy in Britain as being somewhat higher (Lambooy and Renoir 1985) : 3.3 - 15% of GNP in the 1978-80 period (Fase 1984, 83 using a monetary method); and 8.1% of total GNP in 1978 (Frey and Weck-Hanneman 1984, using the unobserved variable method). However, a more recent UK study (Smith 1986) suggests a much smaller UK black economy of between 3-5% of GDP, in contrast to the UK Inland Revenue's own estimates of between 6-8% of GDP for 1979-82.

Similar estimates are available for other EC members (Table 3.24) with Belgium and Italy having the highest levels. For example one source estimates that between 4-7 million workers in Italy (out of total labour force of 22 million) are engaged in the 'economia sommesa'

TABLE 3.24

ESTIMATES OF THE INFORMAL ECONOMY IN OECD COUNTRIES

A) Estimates of the 'Black Sector' as a Percentage of GNP
   Obtained Through Monetary Methods in Selected OECD
   Countries, 1978-1980

	MINIMUM		MAXIMUM
Belgium	15.2		20.8
West-Germany	3.7		27.0
Canada	10.1		23.7
France		6.7	
Ireland		8.0	
Italy		30.0	
The Netherlands	6.3		17.5
Norway	6.3		9.2
Spain		23.0	
United Kingdom	3.3		15.0
USA	2.9		28.0
Sweden		13.2	
Switzerland		3.7	

Source : Fase (1984, 83).

B) Size of the 'Black Sector' as Percentage of Using the
   Unobserved Variable Method GNP for 14 OECD Countries, 1978

Sweden	13.2 (base)		West-Germany	8.3
Belgium	11.5		USA	8.2
Italy	10.5		UK	8.1
The Netherlands	9.2		Ireland	7.0
Norway	9.2 (base)		Spain	6.0
France	8.7		Switzerland	4.5
Canada	8.6		Japan	3.9

Source : Frey and Weck-Hanneman (1984).

(Willatt 1982; see Storey and Johnson 1986, 48). Most studies also conclude that this 'underground' sector has become increasingly significant over the past decade and this together with its apparent concentration in the construction and service industries may suggest that perhaps 10% of all service activities in the Community will go unrecorded by 1990.

Services, Technological Innovation and Location in Europe

Technology and Services - A Framework : A key change element in the development and growth of services in Europe is technological innovation. In the UK study (Howells and Green 1986a, 126-139) a number of points relating to technology and services were made and are reproduced here in a modified form.

The impact of technology on the development of services is profound and has had an interlocking role in the various other structural processes already noted. The introduction of new service activities, new ways in which services are being provided (for example, via new kinds of distributed systems associated with networks or in terms of a substantially altered format) and new forms of service organisation are all having a major influence on the industrial structure of European services. The growing importance of information generation, transformation and transmission and the spread of low cost, widely accessible business telecommunications are just one aspect of technological developments in the service sector.

As in manufacturing a distinction can be made in services between 'product' and 'process' or 'capital' service innovations. The former relate to the creation and establishment of new, and the development of existing, services varying in degree of technological sophistication. The most sophisticated of these service activities are associated with the early stages of the product life cycle (however see later). It is

86

this form of service innovation which is related to the development of totally new service industries and markets at the leading edge of technology.

By contrast process service innovations relate to the impact of technology on the way that a service is generated or delivered. This form of service innovation revolves around the actual generation, manipulation and distribution of a service and the format in which it is delivered. A key distinction (not applicable in manufacturing) to be made here in terms of process innovations and services is between :

 i) the generation and production of services; and
ii) their distribution and delivery.

A third form of innovation in service activity is that of 'disembodied' technological progress associated with improvements in the operation and methods of a service organisation such as changes in work design and the operation of particular functions within services. 'Disembodied' improvements in service activities can be seen in isolation, for example through a new form of service enterprise such as in franchising, but again they are often intimately bound up in product and process developments in service activity. For example the provision of a new service such as electronic publishing may require reorganisation of work practices and the introduction of new specialist ancillary staff. However, it is recognised here that in some senses all service innovations are disembodied since they are, by definition not associated directly with producing or modifying physical objects.

As will be shown, although the distinction between these three main categories of technological innovation may be theoretically useful in our conceptualisation of the effect that technological innovation has on the service sector, in reality trying to properly separate out these different components is extremely difficult and often spurious.

Several studies have rightly acknowledged the major 'trajecting' role that information technology goods and micro-electronics has had on the service sector. However the danger is that technological developments in relation to services are seen only as being exogenous influences emanating from a dynamic manufacturing sector and the service sector at its most crudest acting as a stimulus-response mechanism. Such a view, associated with neo-classical economic theory, also stresses the dominant role of process innovations in cost efficient generation of services. By contrast the indigenous, product related aspects of technological innovation in services tends to be neglected.

It should therefore be acknowledged that service innovations do not have a completely subordinate role to goods innovations (indeed services are vital in the application and role of new technology goods: the most obvious being computer software), that they are often integral to each others development (Krollis 1986) and that there has been a convergence in the development of goods and service innovations. This has been recognised by some large information technology goods manufacturers in the US who have recently moved into joint ventures to provide computerised information services (Table 4.5).

The view that technological innovation is an exogenous factor acting upon the service sector is reflected in the concentration of research interest which has focussed upon the impact of manufactured goods technology on process innovations relating to service generation. This is not to deny the important impact of technology in this form on services, particularly as it relates to capital-labour substitution, growth in labour productivity and labour displacement. However, the importance of indigenous, product related aspects of service innovations, and process innovations relating to the delivery or distribution of services, has largely been ignored by researchers. Thus the employment generating effects of product service innovations, associated with the establishment of new service markets, has almost gone unrecorded. There are some notable exceptions. For example, Barras (1984a, 21-24) notes that the adoption of information technology in financial and business services not only led to quantitative gains in labour productivity and cost reduction but also to equally if not more significant gains in the quality of services. In relation to insurance the information handling capability of large computer systems has allowed for the development of mass, standardised, low cost options which offer more flexible and customer-oriented services (Baran 1985, 171), whilst in banking the advent of networking has led to improved customer facilities the provision of Automatic Telling Machines (ATMs). This has led to the emergence of product service innovations in the form of new packages of customised and packaged financial services covering banking, home loans, insurance and accountancy (Schuster 1986, 155). An example of the complex inter-relationships between goods and service technology and more particularly between product service innovations and process innovations associated with the delivery of services can be seen in the introduction and diffusion of VANS. This is in turn associated with the wider impact of networking on service growth and development (Chapter 5).

Apart from the direct employment impacts of technology, via labour productivity growth which will be discussed below, there are few additional general or theoretical statements that have been made about services and technological innovation (Howells and Green 1986a). An exception to this is a recent statement by Barras (1985; 1986) in a discussion of the diffusion of new technologies in services associated with 'technology push' and 'demand pull' processes. In this discussion he outlines the familiar 'product cycle' theory of innovation, whereby successive waves of innovation in a particular technology shift progressively from product innovations which generate new devices to process innovations which improve the quality or performance of existing devices. However, in relation to services Barras suggests that an opposite process of innovation tends to operate, which can be termed a 'reverse product cycle'. Barras (1985, 3) goes on to note "this implies that the first applications of new technological devices tend to be designed to improve the efficiency of delivery of existing services; subsequent applications then shift the emphasis towards improving the quality of existing services; and finally the process of service innovation reaches the point at which the devices can be said to be generating new service activities. Having identified this reverse product cycle as the mechanism underlying the adoption of technical devices within services, the nature of the 'demand pull' pressures which interact with the technology push pressures behind diffusion now become apparent. For autonomous processes of social and economic development are continuously creating both changing and new service needs, which redefine the functions being served by household and business service activities. These changing functions in turn create demands for improved and new services, which can help to trigger technological innovation and the development of new devices to meet these needs".

This is clearly a novel research statement which needs further verification and has much to commend it. However, for all its statement on demand pull it continues to suggest an emphasis on the dominant rate of exogenous process innovations (associated with developments in information technology) acting upon a passive service sector. As a contrast Lambooy and Tordoir (1985, 5-7) suggest the use of the product life cycle model proper, albeit within a wider context, as an analytical tool in the spatial development of services. Clearly more detailed analytical sectoral research work needs to be undertaken before a more solid theoretical framework is provided on the role of technology in service development.

However, as noted above, one field where there is quite a considerable body of research evidence relates to the impact of technology on service employment, particularly via improvements in labour productivity. This is obviously a key issue as it was suggested that employment growth in services in general, and in the less favoured regions of Europe in particular, was in part associated with lower levels labour productivity growth vis-a-vis other sectors of the economy or other areas. This will be discussed briefly now.

Technology, Employment and Labour Productivity in Services : A series of valuable international research projects have been undertaken which have examined the relation between technology and employment growth in the service (Sleigh et al., 1979; Roda 1980; Robertson, Briggs and Goodchild 1982; Collier 1983; Gershuny and Miles 1983; Barras 1984b; 1985; Petit 1985; Gleave 1985), and more specifically the information (OECD 1981) and office sectors (Steffens 1983), have highlighted low productivity growth in the past as an important factor in the expansion of service jobs. It should be noted, however, that the measurement of labour productivity and output in services is notoriously difficult to undertake and is open to interpretation (Smith 1972; Dewhurst 1985; Fletcher & Snee 1985; d'Alcantara 1986).

There have also been more specific sectoral studies which have analysed the role of technology on employment trends. Thus Hewlett (1985) has looked at the impact of new technology on banking employment in the EC, whilst technology and employment studies in the UK have also examined the banking (Holland 1985), financial (Rajan 1984) and insurance sectors, as well as the professional services (Haywood-Farmer and Nollett 1985), the miscellaneous service sector (Smith 1986), financial services (Mothe 1986) and distributive trades (Brodie 1986). Although all these sectoral studies suggest a number of conflicting trends, the employment impacts of technology are in many cases expected to be profound. For example, the introduction of autobanking has allowed the overall number of bank clerks per customer to be reduced. Indeed Nora and Minc (1978) have suggested that a 30% reduction in possible employment totals in banking has been due to such technological developments. Similarly in relation to word-processor usage Sleigh et al. (1979) have estimated productivity increases of over 100% have been obtainable by office workers.

A key element in many studies is however the differential impact of new technology and automation on occupation and skill sectors (Arnold et al. 1981; Arnold 1983; Doswell 1983; Beven 1984; Barras and Swan 1984b; Child et al. 1984; Crompton and Jones 1984; William and Cowan 1984). The growing dualism in service employment has also been highlighted by the FAST network, although other services employment categories, such as highly skilled information processing and exchange functions, have traditionally been resistant to technological displacement. More crucially here is the effect that these differential impacts will have on the spatial opportunities for service employment growth. As yet apart from a few key studies (Foord and Gillespie 1985; Wood 1986) these spatial issues remain to be fully researched.

As noted in Chapter 2 the gap in the average annual growth rate in labour productivity narrowed between manufacturing and market services during the 1970s and early 1980s. Indeed in Germany and the UK labour productivity increases were greater in services than in manufacturing for the period 1975-7 to 1980-2 (Table 3.25). More recent evidence from Green (1986), however, suggests that with the general slowing in labour productivity in the European Community (EC 7) between the 1960s to early 1970s compared with late 1970s and early 1980s that the catching up by market services on manufacturing in terms of labour productivity growth is now less clearer than it was. Nonetheless the relatively good performance of services in terms of labour productivity growth, at least during the 1970s, is reflected, and in part explained by the growth and relative share of gross fixed investment in market services (Green 1985, 76-7; Green 1986). Thus between 1970-2 and 1980-2 market services share of total gross fixed investment (at 1975 constant prices) rose from 51.5% to 55.4% as an average for EUR6 as a whole (3.9% increase), compared to a decline of manufacturing industry from 20.0% to 17.1% (-2.9% decline). Interestingly Green (1985, 76) suggests that the process of capital widening and deepening in service activities could well be associated with a change in the nature of service producing activities with less emphasis on labour intensive personal services and more on capital intensive services provided to enterprises.

Services, Technological Innovation and Location : As yet very little is known about the level of technological innovativeness and service location in mature industrial economies, particularly in comparison to manufacturing. Nonetheless some recent research work provides a valuable picture of certain aspects of technology and service location. On an aggregate level, Cappellin and Grillenzoni (1983), in their study of the diffusion and specialisation in the location of service activities in Italy, have revealed that there was a positive relationship between the spatial concentration and the degree of novelty

TABLE 3.25

AVERAGE ANNUAL GROWTH RATES OF LABOUR PRODUCTIVITY OVER TWO
FIVE YEAR PERIODS : (i) 1969-71 to 1974-76; (ii) 1975-77 to
1980-82

	MARKET SERVICES		MANUFACTURING INDUSTRY		TOTAL	
	(i)	(ii)	(i)	(ii)	(i)	(ii)
D	2.9	2.8	3.9	2.2	3.3	2.4
F	3.1	1.8	4.1	3.8	3.8	2.5
I	2.3	0.9	3.5	4.0	3.0	2.2
UK	1.7	0.9	2.8	0.6	2.2	0.9
EUR6	2.6[1]	1.8	3.9[1]	2.9	3.2[1]	2.1
USA	1.2[1]	0.6	2.8[1]	1.2	1.2[1]	0.6
Japan	2.6[1]	2.7	5.8[1]	8.4	3.8[1]	4.4

[1] 1970-72 to 1975-77.

Source : Green (1985, 76) compiled from Eurostat and
Commission services.

of various services. Malecki (1977; Brown and Malecki 1977) has examined innovation diffusion in the US between firms within a service sector : banking. His survey analysed the diffusion of computer processing and the spread of credit cards between banks in Ohio and West Virginia, but found that the hierarchical nature of innovation diffusion suggested by Berry (1972) and others did not correspond to the pattern of inter-firm diffusion that was discovered. The spatial pattern of diffusion presented by the studies was one with a strong random element and with adoption appearing to occur firstly in medium-sized rather than large urban centres. Malecki concluded by noting that explanations based on the primary role played by hierarchical or distance dependent information flows among firms in terms of influencing innovation diffusion have been too simplistic.

Bearse's (1978) analysis of the diffusion of US business services as an innovation flow process around the New York City region. He found that the geographic spread of business service activities were contingent on the regional distribution of accessibility to information and specialised resources, in relation to proximity to New York; the size of the professional workforce; and the availability of other inter-mediate services. In addition he found that the location of large firms is a significant influence on the diffusion of business services, a finding which substantiates Pred's (1975) analysis on an interregional basis. Townroe (1986, 85-7) has also suggested that corporate organisation may influence the pattern of service diffusion and technological change (see also Taylor and Hirst 1984; Marshall and Bachtler 1984). More recent work by Jones (1981; 1985) has also provided a detailed picture of the spatial diffusion of retail innovations in the UK. On a more sectoral level van Haselen et al. (1985) have looked at the role of technological change in the banking and insurance sector within the regions of Europe. They noted, for example, a limited decentralising influence in the banking and insurance sectors within the UK out of London, but southern Britain still remained dominant within these sectors.

Work by Daniels (1984) and Foord and Gillespie (1985) has investigated the impact of the introduction of new technologies on the pattern of office employment, the latter examining new technology in the

94

context of the changing geography of women's office jobs. Daniels (1984, 38-9) found that the size of a firm and its ownership status (indigenous/non-indigenous) was less important than the functions performed by the office and the location of organisational control (with provincial offices, which were part of a organisation with a London head office, having higher adoption rates) in terms of the adoption of information processing and telecommunications equipment. Recent work by Foord and Gillespie (1985) has revealed that there were higher levels of take up of the newest information technology in Berkshire than in Tyne and Wear, although the employment effects from this differential adoption rate were by no means direct or simple.

Based on this evidence there is support for the view that there are important core-periphery regional, and to a lesser extent urban size, differences in the level of service innovations and the rate of technology adoption in services in Europe.

Another potentially important development in service locations and work patterns is that of the evolution of 'distance', 'remote' or 'home' working. Distance working allows people to use new information technologies (both hardware and software) to communicate at a distance with their work places and employment. More specifically it is associated with a link between two or more computer terminals via a common-carrier medium (Fogelman 1985). This new form of work pattern is very much in its infancy and it is difficult to forecast its impact on the spatial organisation of service companies (ADR/Empirica/Tavistock Institute 1985). In brief, most commentators in the UK (Huws 1984; Hakim 1984; Holti and Stern 1984) for example, believe that it has considerable potential for changing the organisation and structure of work but has yet only been adopted by a few companies : Rank Xerox, F International, ICL, and Davy Computing Ltd. However, distance working may have important regional implications in terms of increasing the tradeability of information services (Holti and Stern 1985, 129).

In conclusion, although there are several economic advantages in distance working (in particular, in terms of non-salary items such as

office rental and the need for related service functions) particularly
if associated with workers becoming self-employed (based on a minimum
guaranteed consultancy contract), there are also major problems
associated with isolation, loss of worker control and contact (Stern and
Holti 1986, 40) and more indirect, less quantifiable costs. These
latter issues have therefore meant that adoption of distance working has
been, as yet, very limited in the service sector, and that since most
workers have to have face-to-face contact with the office site at least
once a week, most decentralising employment effects will be restricted
to a localised, intra-regional scale. The potential for inter-regional
dispersion of service functions appears to be much more highly
constrained.

Service Location and Regional Development in Europe : Conclusions

Several major fundamental processes have been outlined in this chapter
which are producing profound and often rapid changes in the structure,
industrial organisation and location of service activity in Europe. In
particular what has been striking is that many of these trends have only
recently begun to take off. In part this is a reflection of recent
changes in the regulatory framework both within particular member states
and the Community as a whole (through, for example, the gradual removal
of internal barriers to trade in services). However, it is also a
reflection of key independent structural and technological trends in
services which are beginning to emerge, such as internationalisation and
developments in advanced communication services. Some of these key
issues will be taken up and discussed more fully in Chapters 4 and 5
which outlines the importance and major developments in information
services in the European economy.

In relation to location and regional development the evidence that
has been presented suggests that the evolution of the service sector
over the next ten to fifteen years will militate against the less
favoured, peripheral regions of Europe. Much more detailed research
obviously needs to be carried out to fill out a lot of the tentative
conclusions reached here but the key findings suggest that the major

structural, organisational and technological trends in the EC services will have a centralising effect in inter-regional terms, although on a subregional basis there appear to be few constraints on service activity dispersal. A key implication arising from the analysis is the very strong inter-linkages between the various technological, structural and organisational trends that had been isolated. For example, the increasing concentration of large firm domination of service activity within a wider framework of the evolution of a 'corporate hierarchy' has implications for small and new firm growth, associated with business services linkages and employment structure.

Moreover, although innovation and technological development in service activity would appear to provide a potential liberating role in terms of offering new opportunities in services to less favoured areas, in reality such prospects look bleak. Thus, developments in tele-communications and information technology can allow new forms of service organisation and work practice that are potentially no longer constrained by distance, and offer a new opportunity for more remote, peripheral areas of a national territory. However, in reality inter-woven cumulative organisational and structural constraints appear to be restricting the spread and indigenous development of technological innovations in services.

4 Industrial organisation, technology and location of information services in Europe

Many of the key developments and trends outlined in Chapters 2 and 3 have a fundamental role in the development of new services and techno-logical innovation in the service sector, particularly as it relates to the growth of the information economy. Many of these new service and information markets that are emerging have started from very small bases but are growing at a rapid rate. This has been associated with the substantial growth in the financial and business service markets in western Europe during the 1980s. Within smaller new service market segments growth can be even more rapid. For example, in relation to computerised or electronic information services, Todd and Strasser (1985) estimate that the growth rate of on-line company data, trading support and news/current affairs information services will grow at annual rates of between 20-30% to 60% and over, during the latter half of the 1980s and beyond. This can be seen in the growth in subscribers and revenue of on-line information services (Table 4.1). Similarly, there is a rapid growth in the Value Added Network Service (VANS;

TABLE 4.1

MAJOR ON-LINE INFORMATION SERVICE PROVIDERS IN THE US

PARENT COMPANY	SERVICE PRODUCT	SUBSCRIBERS AT AT 1985 (END)	% CHANGE ON YEAR	1984 REVENUE $M	% CHANGE ON YEAR
H & R Block	Compuserve Information Service	260,000	+ 48.6	9.5a	+ 64
Down Jones & Co.	Down Jones News/Retrieval	235,000	+ 27.0	39.0	+ 19
Mead Corporation	Lexis/Nexis/Medis	180,000	+ 33.3	125.0	+ 31
British Telecom	Dialcom	150,000	+ 87.5	874.0b	+ 4
Quotron Systems	Financial information services	76,665	+ 6.3	175.0	+ 24
Lockheed	Dialog information services	70,000	+ 20.7	59.0c	+ 31
Reuters	Monitor	65,000c/e	+ 21.6	310.0	+ 24
Readers Digest/Control Data	The Source	60,000	+ 1.6	11.7c	+ 35
Equifax	Financial control service	35,485	+ 1.4	60.0	+ 50
ADP Network Services	Bunker Ramo information system	31,500c	+ 5.0	62.1	+ 15
M Dow Jones & Co./Oklahoma Publishing	Telerate	30,000	+ 25.0	114.0	+ 70
Thyssen Bornemisza	Bibliographic Retrieval Services (BRS)	26,000d	+ 116.7	10.0c	+ 25
Dun and Bradstreet Corp.	Dunsprint	24,179	+ 38.2	63.0	+ 26
Knight-Ridder Newspapers	Viewtron	15,000	+ 383.9	n.a.	n.a.
Independent Publications	News Net	11,000	+ 37.5	n.a.	n.a.
Knight-Ridder Newspapers	Commodity new services	10,000c	+ 5.3	40.0c	+ 14
Online Computer Library Centre	OCLC	7,483	+ 20.6	37.0	+ 19
IP Sharp Associates	Sharp/APL	6,700	+ 11.7	9.49	+ 9
General Electric	Mark III	6,090	−	8.0c	+ 18
Knight-Ridder Newspapers	Vu/Text information services	2,360	+ 145.8	n.a.	n.a.

a Excludes electronic and computer conferencing. b Communications operations and information service division. c Estimate.
d As of September 1 1985. e Worldwide.

Source : based on Nachoma

defined as services publicly accessible which add extra value to the functions of the basic telephone network in terms of offering more than the basic provision of voice or data by in some way manipulating the information on behalf of the owner such as the reformatting; storage and retrieval of information; Bleazard 1986, 3-7) market in Europe. Thus in terms of annual revenue the Western European market was worth only $270 million in 1984 (Frost and Sullivan Inc 1985); however by 1986 the market had risen to $900 million and by 1991 this has been forecasted to grow to over $4.8 billion in constant dollar terms, representing a real annual average growth rate of nearly 40% (Frost and Sullivan Inc. 1987).

This explosion in new service and information markets is also evident in employment growth. For example, in the UK the highest growth rates in employment during the 1980s are in fields on a within Activity Heading (4 digit) basis which are not yet properly classified, i.e. in 'other', 'ancillary' or 'not elsewhere specified (NES)' categories associated with producer related, information sectors (Howells and Green 1986a, 219). It is also reflected in the growth of information occupations (Chapter 2; Table 4.2).

The fundamental importance of the information services in the European Community has been highlighted in a recent report by the European Commission (1985, 6) which noted the key position that information now plays in modern society in particular in relation to trade and industry.

In particular, the focus on information services in this chapter is for five main reasons :

(1) The key and expanding role that information plays in the increasingly integrated system of trade and industry.

(2) The rapid and sustained growth of advanced information sectors in the European Community.

TABLE 4.2

INFORMATION OCCUPATIONS AS PERCENTAGES OF ALL ECONOMICALLY ACTIVE

	1951	1961	1971	1975	1981	1982
Australia			39.4			
Austria	18.0	22.0	28.5	32.2a		
Canada	29.4	34.2	39.9			
Denmark				30.4b		
Finland	12.6i	17.3g	22.1c	27.5b		
France	30.3d	24.1e	28.5f	32.1		
Germany	18.3i	23.4	29.3c	32.8a	33.5	34.8
Japan	17.9g	22.2h	25.4c	29.6		
New Zealand				29.4a	39.8	
Norway				20.8	22.9	
Sweden	26.0g	28.7h	32.6c	34.9	36.1b	
United Kingdom	26.7	32.1	35.6		41.0	
United States	30.7i	34.7g	41.1c		45.8b	

a.	1976	b.	1980	c.	1970	d.	1954
e.	1962	f.	1968	g.	1960	h.	1965
i.	1950						

Source : OECD (1985a, 4.)

(3) The pivotal and integral role that information services play in technological and structural trends in the economy of Europe. In particular information services are important in the role they play as the source and indicators of technological innovation in the economy.

(4) The important function that information services play in the growth and development of firms (see, for example, Antonelli 1984; 1985); in particular how industrial organisations respond and adjust to changes in their external operating environments.

(5) The increasing 'transportability' of information services associated with (Rada 1984, 3; Feketekuty 1986, 590) the development of computer-communication networks.

The rest of the chapter will analyse the key elements in the growth and development of information services, in particularly concentrating on the new and and technologically advanced information service segments, and their spatial impact.

The Emergence, Growth and Economic Structure of Information Services

The Commodification, and Externalisation of Information: Information services are not only increasingly important as a constituent of the overall economy but are also becoming significant in their own right as they are commodified and develop into separate sectors of their own. As such a fundamental factor in the growth and development of information services in Europe has been the commodification of information services and the evolution of information handling units within large industrial firms (associated with internalisation and externalisation processes; Chapter 3) and related issues associated with the internal economy of the firm.

Information service products have often emerged from information activities that have been undertaken within industrial firms. These activities, set within internal corporate divisions or departments such as information processing, design accountancy or marketing have until recently largely gone uncosted or even unrecorded within the firm. Thus even though internal corporate departments might be run as cost centres (where the budget and activities of the centre are planned at the beginning of each budget period for individual sub-units or sub-activities) there may be little or no cost control or any idea of the transfer price of the information commodity involved (Radner 1986, 6).

It may be argued that the reason why information service activities in industry have grown so rapidly in recent years is for the very reason that they have gone uncosted and therefore remained a hidden element in the corporate economy. Thus inputs of information services have increased in order to reduce the use, and therefore cost, of non-information inputs which are more readily costed in the manufacturing or service process. The net result of this may be that the firm increases its overall input/cost structure because the uncosted information service inputs are in fact more expensive to use than other cost-quantifiable inputs. More directly in terms of employment costs other changes in both manufacturing and service industries, (in particular increases in capital intensity leading to savings of labour in direct manual occupations), have increased demands in more specialised, higher grade information occupations (and also in some indirect manual jobs) such as maintenance.

Overall the cost of producing and using information services in a firm, their transfer to other parts of the firm, and their pricing are often ill-defined and badly organised. Thus the price structures that are evolved within the firm (if they have any at all) for these information services may have little or no bearing to their actual cost of generation or distribution, but rather are more heavily influenced by other factors associated, for example, with transfer pricing policies. Indeed in this respect information services allow even more scope for transfer pricing than manufactured goods (Green 1981, 223) simply because there are fewer open markets, normal prices or identical products. Thus contributions to 'central overheads' or 'head office charges' by branch establishments for services may have little or no bearing to either their cost or use (Green 1981, 223 and 231).

Related to this the impact of the trend towards the greater generation and utilisation of information services within industry has until recently been confined internally within the firm. As such the internal function providing the service simply grew pro-rata according to the increasing demand and use of the company of that service. However, increasingly information services have been sold as a commodity to other firms and this has been associated with the increasing externalisation of service products and operations selling service products noted earlier (Chapter 3). As such inter-company service trade flows, particularly in information services, has been growing and associated with this the service operations of companies have become more open. Moreover, as firms have found that their internal service functions have external markets which are highly valued and in great demand, this in turn has meant that these same firms have now started to reassess their own demand and cost of these internal services themselves.

As part of this trend, for example, a growing number of industrial companies are becoming aware of the value of their information data bases. In the UK for example, a major consumer electrical manufacturing company, Thorn-EMI, which has developed a strong retail rental chain, realised that it had a detailed information database of 3.5 million customers which covered names, addresses, demographic, financial and other details. On this basis it is using its database to market direct mail and financial services and hopes to sell its database to other companies offering direct-mail and personal services. Similarly, another UK company, Marks and Spencer which has now a large credit card operation of 1.3 million customers, hopes to use this customer information base to market other financial services. A final example is that of IBM which as part of its drive towards developing its information service business, is offering customers access to its own database via a special computer-communication network, Info-Express; so far it has attracted 700 corporate customers.

The implications of all these trends are having a major and fundamental impact on the development of services in mature industrial economies. Not only are industrial companies selling services produced by their internal information functions operating as cost/profit centres or subsidiary companies, but they can in turn also assess whether some other internal service functions (such as computer services) can be more efficiently undertaken by outside firms for them. This has a number of implications:

(1) more <u>operators</u> are appearing in these new information service niches;

(2) a <u>market</u> is developing for these services;

(3) associated with this a clearer <u>value</u> is being established in cost/price terms for such services; and finally,

(4) a more distinct <u>industry sector</u> pattern is developing out of these cumulative processes. Some examples of these new, emergent sectors are electronic information services, market research, business consultancy, and employee recruitment.

Perhaps the best example, presented below, of such an emerging information industry is computer services (Chapter 3; Howells 1987b) in part because of it one of the oldest 'new' information service sectors. It also highlights specific sector factors associated with the emergence of information services as well as the effect that changes on the legislative framework may have on encouraging such developments.

The emergence of these in-house computer service operations are very much associated with the changing nature of data processing within industrial organisations. Thus the highly specific character of the expertise necessary to use computer systems and the centralised nature of the hardware infrastructure meant that specialist data processing departments were created during the 1960s and early 1970s. By the early 1980s because of the sheer growth and all-embracing nature of computing meant that by the late 1970s most data processing departments held a key position in organisations. Moreover, data processing departments often became independent of the general strategies or constraints of the organisation and were able to define their own aims and how they could be achieved with little or no interference from the rest of the organisation. In order to reassert economic constraints and targets an increasing number of organisations decided to hive off the department altogether into a separate profit centre (OECD 1985b, 49).

The importance of, and development out of, in-house production of computer software and services also stems out of the nature of software production in terms of the dominance of the operation and maintenance stage in the life cycle of software (OECD 1985b, 29). A study in 1978 by the National Bureau of Standards found that 67% of all investment in software over its life-cycle was in maintenance compared with only 33% for creating and testing it. Similarly in 1980 it was estimated that 75% of all US computing resources were absorbed by maintenance activities and that this was reflected in the fact that 73% of US data processing personnel were located in user firms in 1979 (OECD 1985a, 35), whilst in the European Community this was estimated to be 80% (Perring 1983, 119). Most of the expenditure in software is therefore in maintenance related to having to revise software programs due to alterations in its external environment associated with new user requirements and so on. Much of this work is undertaken by the data processing staff of the end-user rather than the original supplier of the software. As such much of the expenditure on software does not go to the originator but remains with the user which in turn develops its own skills and expertise in terms of the design, maintenance and operation of the software it uses.

A further factor in this process in the computer service industry which has not yet been mentioned is the impact of changes in the legislatory framework. From the early 1960s computer manufacturers have 'bundled' their systems software into the price of the hardware with no separate invoicing. However, this practice was drastically changed in 1969 when the United States Department of Justice required IBM to invoice its hardware and software separately (Sobel 1981, 261-266). This 'unbundling' had a major impact on stimulating the growth of independent software suppliers and the packaged software market.

A final element here in the actual development of the commodification, selling and the wider externalisation process in information services has been the growth of inter-organisational computer-communication networks (Chapter 5), which has allowed certain information and data services to be more easily and cheaply provided on an inter-organisational basis (Estrin 1985; 1986). Thus inter-organisational trade flows in information services has been encouraged; companies in terms of information generation and provision have become more open which in turn has allowed more information sharing and exchange. The evaluation and development of corporate computer and information system have therefore been closely associated with the development of computer-communication networking. Corporate computer-communication systems have moved on from linking geographically separate parts of large multi-site, multinational companies (Table 4.3) to developing links with other related companies on a horizontal, intra-industry basis (for example, in banking CHAPS in the UK or SWIFT as an international level) and vertical, inter-industry basis (for example in the automative industry, ODETTE in Europe and ZENGIN in Japan : Ministry of Posts and Telecommunications 1987). Such developments have also meant an ever widening of the geographical field of contact from local through to regional, national and international networking arrangements. Such developments may encourage large companies to specialise on those information activities that they are best at and exporting them to other companies whilst importing other types of information activities which they are less well placed (in terms of cost, efficiency, degree of specialisation or frequency of use) to generate or service. It may also encourage certain service companies with large international computer-communication networks to diversify in terms of the services they provide (see Richardson 1986, 388; Desiata 1986, 92). This is because the fixed costs of the network are high, but the marginal costs of putting more traffic through the network are low. Further, companies with their own private communication networks will start to sell their excess capacity on their networks to other companies and therefore will start competing with the telecommunication companies themselves. Thus in the US, Sears Roebuck, Merrill Lynch and McDonell Douglas are now reselling their excess communication capacity (Irwin and Merenda 1987, 18).

TABLE 4.3

TRENDS IN CORPORATE COMPUTER AND INFORMATION NETWORKING

Stage	Computer/ Information System	Network System	Scale of Standardisation
1.	In-house batch processing		
2.	In-house on-line system	Private LAN/WAN	In-house
3.	Dedicated, jointly operated system.	Dedicated VAN	
		Horizontal (Intra-Industry)	Within corporate groupings
		Vertical (Inter-Industry)	Within industry groups
4.	Domestic network	General Purpose VAN	National agreement
5.	International network	Internal VANS	International agreement

LAN — Local Area Network   VAN — Value Added Network
WAN — Wide Area Network

Source: based on JIPDEC (1987, 68).

<u>New Entrants, Concentration and Growth in New Information Service</u>: There are a number of key processes at work in the emergence and development of new information services. On the demand side is the actual growth in information activity within the economy which is reflected in the internal growth of information activities within the firm as well as the growth in bought-in information services from other companies. On the supply-side this is reflected in the creation of new information service enterprise via the commodification and externalisation of information service activities from within firms, as well as the more specific diversification and acquisition strategies of large corporations seeking to develop information service components within their portfolio of activities. There are also the creation of totally new firms involved in information service activities. Finally there has been the role of technological change itself associated with the development of computer technology. Thus storage and processing capacity increased dramatically during the 1950s, 1960s and more particularly 1970s, whilst unit costs were dramatically lowered (see for example Fertig 1985, 82–89). For example, calculation costs per 100,000 fell from $1.26 in 1952 to $0.0025 in 1980, whilst on-line storage costs fell from $1.43 million per 10 bytes in 1973 to only $0.392 million by 1981. As such third and fourth generation computers have become available they have greatly increased capacity and lowered storage and processing costs for database producers and electronic information services (CSP International Ltd 1986, 23).

Chapter 3 indicated that although new and small firms represented a buoyant sector in services this was within an overall market framework in the Community of increasing concentration of ownership. Thus a major dynamic element within the service sector is coming from existing large and medium sized industrial corporations who are moving into key service markets through diversification or externalisation. Subsequently some of these service operations are spun out to form independent companies (for example, through management buy-outs) or acquired and absorbed into other large service or conglomerate corporations. It is presented here that this pattern is also true within information services, for example, the UK computer service industry has a very high rate of new firm formation (birth rate of 242.2% between 1980 and 1983) but within the context of an elite set of computer service companies operating in the UK market which appear to be holding onto and strengthening their market position. As such most of the key actors in the UK computer services industry have not arisen out of new firms which have experienced rapid growth over the last ten or fifteen years (the 'high tech' dream) but have entered via internalisation/externalisation processes in conjunction with diversification and acquisition trends (Table 4.4).

The process of externalisation is also true in other information service markets. In financial information services, for example, those leading-edge financial institutions which developed their own internal on-line information systems rapidly become information suppliers to third parties in their own right (Todd and Strasser 1985).

The wide variety of participants and entrants to many of these new information sector can be seen in the developing electronic information service market and out-markets. These actors range from :

(1) Existing, traditional press and publishing companies who have found their existing databases and/or their marketing strength extremely valuable in terms of developing an operation on their own or in collaboration with other participants (Table 4.5). Thus in the US Reader's Digest, Oklahoma Publishing, Knight-Ridder Newspapers and Independent Publications are all key participants in the US on-line information market (Table 4.1). In Italy many publishing companies and news agencies are also developing their own information database services (Lazzari 1985, 13).

(2) Large industrial corporations which have developed databases (Gurnsey 1986, 101) for their own use, have subsequently found them marketable to other companies. Lockheed in the US, for example, was the first on-line database to be launched in 1972 which was based on essentially bibliographic data for science and technology.

TABLE 4.4

EXTERNALISATION, DIVERSIFICATION AND ACQUISITION ACTIVITY IN UK COMPUTER SERVICES :
SOME EXAMPLES

NAME OF PARENT COMPANY	MAIN OPERATIONAL SECTOR OF COMPANY	NAME OF COMPUTER SERVICE OPERATION
1. Computer Service Operations Remaining Embedded in Large Corporations (Externalisation i) - iii))		
National Westminster Bank	Banking	Centre-File Ltd.
British Petroleum	Oil	Scicon International
British Coal	Coal	Compower
2. Computer Services Operations Sold Off by Large Corporations (Externalisation iv))*		
Unilever plc	Food/household products	Unilever Computer Services Ltd.
P & O	Sea transport	P & O Computer Services
Heron International	Financial & Property Services	First Computer
Rover Group	Cars	Istel (Computer and information systems)
3. Acquisition : Purchase of Computer Service Operations by Large Corporations		
Reed International	Publishing and other activities	Hooper Systems

* Refer to Table 3.4

Source : Howells (1987b)

TABLE 4.5

MAJOR EUROPEAN PUBLISHING COMPANIES MOVING INTO

ON-LINE INFORMATION, BROADCASTING AND COMMUNICATION RELATED SERVICES

NAME	NATIONALITY	EXAMPLES OF SUBSIDIARY COMPANIES
Arnoldo Mondadon Editore SpA	Italy	Editoriale Televisiva R4 Srl (SEDIT) Supernova Ediziono Srl MKT Italia SpA.
Elsevier NV	Netherlands	Elsevier - IRCS Ltd (UK) Elsevier Business PCSS Inc (US) European Video Corp.BV Congressional Information Service Inc (US) Elsevier Scientific Publishers Ireland Ltd (IRE)
Verenigde Nederlondse Vitgeversbedrijen BV	Netherlands	Computing Publications Ltd (UK) Filmnet Aboumee Televisie Nedeland (BV) Publincinformatica SA (E) Educational Network BV
Hachette SA	France	Channel 80 40% stake in the Europe-1 broadcasting group Saarbach Gimbh (FRG) Seymour Press Group Ltd (UK)
Bertelsmann AG	FR Germany	A-Z Direct Marketing BU (NL) Bertelsmann Datenbankdienst Gmbh (FRG) Bertelsmann Information Services Gmbh (FRG) Bantam Books Inc (US)
British Printing and Communication Corporation plc/ Pergamon Press Ltd*	UK	Pergamon Infoline Ltd Delphin Verlog Gmbh (FRG) Milthorp Srl (I) British Cable Services Ltd Selec TV Plc Mirror Group Newspapers Ltd
Gutenburghus Group	Denmark	Select Video Ltd (UK) Deutsche Egmant Verwaltungs Gmbh (FRG) Nederlandse Egmant BU (NL)

* Registered in Lichtenstein

(3) <u>Telecommunication companies</u> have also moved into electronic information services via being host operators or through information processing facilities. Examples of this are British Telecom's Prestel and Dialcom services and the Bundesposts Bildschirmtext service or on a wider basis the establishment by NTT of a data communication and information processing division (including a joint venture with Kyodo News Service on an on-line news and current affairs service).

(4) <u>Information technology goods manufacturers</u>, as we have seen (Table 4.6), have also moved into information service market sectors either in partnership with major information service companies or on their own. A European example of this is Olivetti's acquisition (in conjunction with Institute Bancario San Paolodi Torino and Cassa di Risparmiode) of Pitagora Information Services, a financial services provider.

(5) <u>New and small information service niche companies</u> associated with specialist and bespoke electronic or desk-top publishing often in conjunction with consultancy or typesetting operations.

(6) <u>Institutes or professional societies</u> which have developed databases covering scientific or technical data for their benefit.

(7) <u>Government or Public Sector Agencies</u> which produce databases arising out of direct government policy to stimulate advanced information services or indirectly as supporting technical infrastructure to wider science and technology programmes. Examples of this can be seen in Germany on a national (Federal government) basis with the DITR database associated with the German Information Centre for Technical Rules and in a national/regional basis with a joint Federal/Lander government joint venture, F12 Chennie Gambh, which provides a chemical database. Equally on a European level there is the successful development of the Diane network ran by the Community as well as indirect pan-European ventures, such as, the science and technical database operation (ESA-IRS) of the European Space Agency.

TABLE 4.6

FINANCIAL AND BUSINESS INFORMATION SERVICES : LINKS WITH INFORMATION TECHNOLOGY GOODS CORPORATIONS

INFORMATION TECHNOLOGY GOODS MANUFACTURERS	VENTURE PARTICIPANT	SERVICE	MARKET:	COMMENTS
AT & T	Quotron Systems	Financial information system	Stockbrokers/analysts	Quotron acquired by Citibank in 1986.
AT & T	Chemical Bank Bank of America Time Magazine	Home banking and brokerage videotext service ('Pronto')	Individual/small businesses	–
IBM	Merrill Lynch	Electronic financial information service (IMNET).	Stockbrokers/analysts	Abandoned in 1987
IBM	Sears Roeback & Co. CBS	Videotext home information service ('Trintex')	Individual/businesses	CBS withdrawn from project
Wang	(No participant)	Electronic financial information service (Shark)	Stockbrokers/analysts	–
Instinet (US supplier of a computerised system for funding shares)	Reuters	Automated trading system covering stock markets worldwide	Stockbrokers/analysts	Instinet acquired by Reuters in 1986
ITT	(No participant)	Dialcom	Electronic mail	Dialcom acquired from ITT by British Telecom in 1985.

Source : Updated from Howells and Green (1986a, 129).

114

7. Other companies involved in related service or technology activities, such as computer software firms.

There are therefore a wide range of participants and entrants into a new information market such as on-line services. However, before analysing some of the key structural trends in such newly emerging industries it is useful to provide an outline of some of the key features of the economic framework of such embryonic service industries. A brief review of the structure and function of electronic information services, using valuable material from a recent publication by CSP International Ltd (1986), is therefore presented as an illustration of such a new service sector.

Economic Structure of Electronic Information Services: In outlining the structure of electronic information services it is important to note the four main phases of information production and provision that are commonly recognised.

(1) The organisation of the raw data by editing or formatting.

(2) The production of information by printing.

(3) The distribution of information, which may cover:
     i)   the physical delivery of hard copy (eg. newspaper);
     ii) delivery via broadcasting media (eg. radio or television);
     iii) delivery via telecommunication networks (eg. on-line database services); and
     iv) physical delivery of electronically recorded information (eg. optical discs ie. CD-ROMs)

(4) The access by the consumer/user, involving:
     i)   manual searching/selective (eg. book); and
     ii) searching by electronic means (eg. videotex). Access in this case may be stored on tape or disc, or long distance over telecommunication links.

Electronic information services (defined as a product or service which provides electronically selected and edited information through electronic media linking user and information source) are those products

115

or services which in the organisation (1) and access (4) stages are electronically based, although if these stages are electronic then the other two are nearly always as well (CSP International 1986, 4). An exception to this is the rapidly growing optical disc, CD-ROM, publishing activity in which distribution (stage 3) is via the postal system.

However, in addition to the four basic elements of electronic information production and provision are the main roles in the industry. These cover (CSP International Ltd 1986, 7)*:

(1) The information provider who produces the information, owns the copyright and makes it available for electronic distribution.

(2) The publisher or 'packager' compiles and 'packages' the information to create an on-line service with a marketable product identity. The packager adds value by applying relevant format and presentation standards to enhance the image of the product/service and thereby achieve consumer acceptance, and may also bring together databases on related topics from diverse sources under the umbrella of a single service.

(3) The host organisation takes the content packaged by one or more information providers or packagers and offers computer facilities which enable user to access this information.

It should be noted that alternative role structures for electronic information services have been presented although they are similar in nature. Thus Information Dynamics (1986, 52) have four basic supply categories: information providers; information vendors; communication service suppliers and the customer delivery vehicle. Williams (1984) presents a more detailed supply chain for electronic database use consisting of nine stages: generator of information; primary publisher; secondary publisher; on-line vendor; intermediary system; broker; intermediary searcher; user (end user); ultimate user. The basic differences in these classifications of stages in electronic information services stem from the somewhat differences in perspective: ie. roles (CSP International), supply structure (Information Dynamics) and usage chain (Williams).

FIGURE 4.1

STRUCTURE OF THE ELECTRONIC INFORMATION SERVICES INDUSTRY

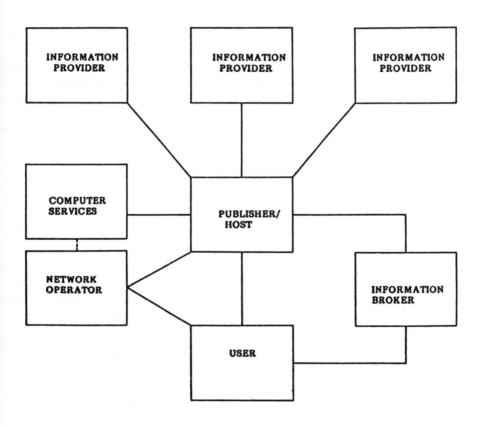

Source: CSP International Ltd (1986, 8).

(4) The network operator provides telecommunications links between the end user and the system operator and, where necessary, between information providers, packagers and host organisations.

(5) Information brokers provide end users with specially designed information products or items particularly suited to their needs. The broker adds value by searching existing products or services to provide information tailored in form and content as closely as possible to end users' needs.

Figure 4.1 outlines the interaction between the different role agents although often these roles are provided by one organisation, in particular in relation to packaging and hosting; similarly network operators often perform the role of information brokers. As such the key distinctions are between information provision, hosting network operations although increasingly companies are seeking to perform all three functions. It has been estimated that a representative split in the level of value added between roles (typically there would be no packager) for an on-line database application would be 15% for the information provider, 75% for the host organisation and 10% for the network greater (CSP International 1986, 26-7; see also Information Dynamics Ltd, 1986 for more detailed information relating to tariff and cost structures). Not surprisingly established on-line information service operators have sought to establish their host operator role although increasing competition and the increasing use of microcomputers and downloading by users has tended to limit the profitability of such operations. Most operators have sought to take over information providers to acquire consumer loyalty that many information providers have built up through close customer relationships. Equally however, information provides (such as McGraw Hill acquiring Data Resources Inc) have sought to move into what they see as the pivotal role of a host operator.

Moving onto the cost structures of the industry (Table 4.7) both information providers and packages have low capital costs compared to host operators and more particularly network operators. In relation to

TABLE 4.7

ECONOMIC COST STRUCTURES OF ELECTRONIC INFORMATION SERVICES

| ROLE | CAPITAL COSTS | OPERATING COSTS | |
		ECONOMIES OF SCOPE	ECONOMIES OF SCALE
Information Providers	Low	–	***
Packagers	Low	***	***
Host Operators	High	**	*/**
Network Operators	High	***	***

- –   Marginal
- *   Limited
- **   Significant
- *** Very significant

operating costs which can be affected by economies of scope and economies of scale there are important variations. For information providers these are substantial economies of scale for database providers, although in terms of economies of scope associated with moving into a new field of information, there are few, benefits unless it is in a closely related field. In terms of providers substantial economies of scale and scope prevail as fixed costs are spread a great number of users and information products. For host operators economies of scope are significant although there appear to be limits in terms of service quality and response times, whilst economies of scale are more limited in scope. Finally for network operators there are highly significant economies of scope and scale via increased network utilisation.

Competition and Growth in New Information Services: Although new and small on-line information service firms appear to be still finding or creating niche markets as information packages and providers, major information service producers are seeking to control the whole market via the pivotal and dominant positions as host operators and through major acquisition programmes. Moreover although information provision and packaging does have low entry barriers in terms of capital costs, the important scale economies (and scope economies in packaging) do mean that existing, large-scale information provision/packaging operations have a major competitive advantage. As we have seen the high capital costs in combination with the significant scale and scope economies of host and network operators effectively precludes SME competition. Not surprisingly the high entry barriers and scale economies in the key activities of on-line information services is causing concern to the Commission as a major emerging sector of the European industrial economy (Commission of the European Communities 1985,7; although interestingly, Cullen (1986, 305) suggests that it is high variable costs that have been a major obstacle to electronic information service market development).

Related to this on the user side are the costs of gaining entry and developing computer-communication networks. For example, access to the

Odette (Organisation for Data Exchange by Teletransmission in Europe) 'mutual' or 'league' delivery system based in London and developed by the European automobile industry for on-line messages, billing and accounts, costs as much as £6,000 in terms of computer hardware, software and services to join. Most small car component suppliers are not able to afford to gain access to this network and therefore risk being increasingly left out of component contracts. As such there is concern that small and medium-sized companies are likely to be excluded from using networks both as a system to serve internal corporate needs as well as externally in terms of selling information services to customers.

Not surprisingly in order to achieve market dominance and gain economies of scale in the computerised or on-line information services there have been a series of acquisitions, on both a product and geographical level, and moves towards vertical integration of the market. Thus McGraw-Hill Publishing Company bought Data Resources Inc. (DRI), an on-line information service company, in 1979 and more recently Systemetrics Inc, a medical database service, and Gnostic Concepts Inc., a market research company. Similarly, Reuters Holdings plc, the UK based international news organisation, has not only acquired Finsbury Data Services, a UK financial database provider in 1986, but has also acquired or taken part share in three information technology (IT) hardware manufacturers; Rich Inc., IDR Inc. and Instinet Corporation, all in the last few years (Table 4.8). It is now in the process of acquiring I.P. Sharp, a major Canadian on-line information provider, further strengthening its position in financial and business information services.

Reuters has now become the world's largest on-line information service provider with a communication network which has 276 international circuits connecting 118 different countries (Chapter 5). Reuters in particular has come to dominate the world information market financial trading systems. Similarly, goods manufacturers have sought to be involved in joint partnerships with companies offering information services (Table 4.6). This has enabled them to move into a high value

TABLE 4.8

SELECTED SUBSIDIARY AND RELATED COMPANIES HELD BY REUTERS HOLDINGS PLC

	COUNTRY OF INCORPORATION	PRINCIPAL AREA OF OPERATION	% OF EQUITY MOVES HELD	NATURE OF BUSINESS
Subsidiaries :				
Reuters Ltd	UK	Worldwide	98.8	International news organisation
Rich Inc.	USA	USA	100.0	Designer of communications systems for financial funding rooms
IDR Inc.	USA	USA	100.0	Manufacturer of terminal equipment
Visinews Ltd	UK	Worldwide	55.0	Television new agency
Finsbury Data Services	UK	Worldwide	100.0	Historic financial database provider
Instinet Group	USA	Worldwide	100.0	Computer based system for automated equity trading
I.P. Sharp	Canadian	Worldwide	100.0	Interactive financial and business services
Related Companies :				
AAP Reuters Communications Pty Ltd	Australia	Australia	44.2	Provision of communication and research facilities
Visinews Ltd	UK	Europe	33.1	Electronic transmission of news pictures
Societe Francaise Pour l'Information SA	France	Europe	50.0	News and business research

added business and also allowed them to create a captive market for their machines with customers using the information system. At present competition in the electronic information service market, particularly in relation to financial information services, has become increasingly aggressive as the market has matured and existing suppliers, such as Reuters and Dow Jones, have established sophisticated and extensive information network systems. In order to compete with these existing companies new market entrants have been faced with steadily rising costs in terms of setting up an information service network which can offer comparable services to those that already exist. As a consequence entry barriers have risen substantially with much larger capital sums now required to set up the information system and consequently much longer payback periods. Thus IBM and Merrill Lynch have decided to wind up their IMNET venture, and CBS has pulled out of the Trintex videotex information service. However, new participants are still appearing: Citibank acquired Quotron Systems, Wang is setting up 'Shark' a financial information service, whilst British Telecom has acquired Dialcom the international electronic mail service, from ITT.

European Information Service Trade and Investment

Although the EC may be in a relatively strong position in terms of international service trade and investment (Chapter 3) in relation to new and innovative information services, which are exhibiting high rates of expansion, the European Community's position would appear to be weak. Thus there is concern that are few truly international European service companies operating in the new fields of computer, business and information related service sectors, such as management consultancy, advertising, computer software, accountancy and electronic information services.

In relation to this, it has recently been estimated that over half of the on-line financial information service markets in Europe (acknowledged as the lead growth sector in information services) was controlled by non-European, mainly US suppliers (Todd and Strasser, 1985, 8; see also UK Association of Data Producers 1986). There are now one or two major European multinational electronic information companies which have emerged as international competitors. Most notably these include: Reuters (UK) based in London; Bertelsmann Information Service Gmbh (FRG) based in Munich; Telesystems Questel (Fr) based in Paris and owned by DGT and France Cable et Radio; and WEF Associates AG (Swiss) which owns Wharton Economics (although the operation is based in Philadelphia). However foreign, particularly US companies have now come to dominate both the US and European electronic information service markets (Table 4.9; see also Ducker 1985; Sema-Metra 1986). Thus Lazzari (1985, 1) in the Italian context has commented on the fact that Italy's on-line information market is characterised by a strong presence of foreign data-base produces. Some US companies have followed aggressive growth and acquisition strategies both in the US and Europe. For example, Dunn & Bradstreet Corporation has acquired A.C. Neilsen's, a US market survey company with substantial operations in Europe for $1.1 billion; Schimmelpfeng, a West German company; and Datastream plc, a UK-based company. Similarly IMS Incorporated has established a strong network of subsidiary operations in Europe providing medical and demographic information systems (Table 3.15). Other US companies however have sought to develop partnerships with European companies (Table 4.10). A number of US companies have seen the potential benefits associated with a link-up with a key European partner. In particular IBM has sought links with European companies, such as German (Hemdelsblatt and Online) and UK (Extel) electronic publishers and Fiat in a stock control network (Table 4.10); although its potentially major link-up with British Telecom in a joint venture network, code-named 'Jove', was blocked by the UK government on the grounds of competitive policy. The 'Jove' network, associated with Electronic Funds Transfer at the Point-of-Sale (EFTPOS) to banks, retailers and credit and companies, would have created a data communications network which would have enhanced the dominant UK market position of both IBM and British Telecom (Information Dynamics Ltd 1986, 68; Bruce, Cunard and Director 1986, 441-2). It also raised more fundamental regulatory issues in the UK of what was a basic conveyancing network, based on a 'Managed Data Network' (MDN) and what was a Value Added Network Service (VANS).

TABLE 4.9

MARKET SHARE OF MAJOR ON-LINE INFORMATION SERVICE COMPANIES IN EUROPE AND THE UNITED STATES, 1985

DATA BASE CATEGORY	EUROPE		UNITED STATES	
Economic/Econometric Analysis	Data Resources Inc - (DRI) (McGraw-Hill)	: 40%	DRI (McGraw-Hill)	: 45%
	Interactive Data Corp - (IDC) Chase (Manhattan)	: 30%	IDC (Chase Manhattan)	: 30%
Financial & Stock Exchange	Reuters	: 90%	Quotron	: 50%
			Telerate (Dow Jones & Co Inc.)	: 20%
			Bunker Ramo (ADP Network Services)	: 20%
Financial Analysis	Dun & Bradstreet Corp (via Datastream)	: 40%	ADP	: 40%
			Control Data	: 10%
			I.P. Sharp Associates	: 10%
Company Information Personal Credit	Dun & Bradstreet Corp	: 40%	Dun & Bradstreet Corp	: 80%
			TRW	: 30% *
Marketing & Demo- graphic Trends	Dun & Bradstreet Corp (via AC Nielsen)	: 20%	Dun & Bradstreet Corp (via AC Nielsen)	: 20%
	IMS Inc		IMS Inc.	: 10%
			PRC REALTY	: 5%
Law & Accountancy	Mead Corp	: 30%	Mead Corp	: 80%
News & Current Affairs	Mead Corp	: 45%	Mead Corp	: 40%
	Lockheed Corp	: 30%	Lockheed Corp	: 35%
Reference	SDC	: 20%	Online Computer Library Centre (OCLC)	: 25%
	IRS (ASE)	: 10%	Burroughs (SDC)	: 25%
	Telesystems Questel (DGT and France Cable et Radio)	: 5%	Thyssen – Bornemisza (BRS)	: 10%

* Estimate

Source : Data from INFOTECTURE

TABLE 4.10

US CORPORATE LINKS WITH EUROPEAN INFORMATION SERVICE PROVIDERS AND DISTRIBUTORS

US COMPANY	EC COMPANY (NATIONALITY)	INFORMATION PRODUCT/VENTURE
Mead Data Control	Teleconsult (France) (part of the Gamma group)	LEXIS legal database
IBM	Handelsblatt (economic publisher) (FR Germany) + Online Gmbh (FR Germany)	Economic information systems
IBM	Fiat (Italy)	Stock control network – In. Te. SA.
IBM	Extel (UK)	Joint VAN service
McDonnell Douglas	British Telecom (UK)	Joint VAN service – now abandoned
Pacific Telesis	acquired : Kensington Data System (UK)	VAN services
American Express + Citicorps (via Diners Club)	Olivetti + STET + Sixcom (Italy)	VAN service company – SEVA
Geisco (owned by General Electric)	ICL (UK)	VANS service company – INS providing electronic transactions and information exchange.
BRS*	Elsevier (NL)	Electronic publishing – IRCS Medical Science Journal

* based in the US under Indian Head Inc. but ultimately owned by Thyssen – Bornemisza NV, Cyrcao (Netherlands Antilles).

A number of reasons have been put forward of why there has been a lag in the development of advanced information service companies and trade in Europe including lack of common standards and the role of multilingualism (Middleton 1985). However, one factor which is associated with this lack of development in the more advanced, innovative sectors of information service is that of internal barriers in the EC market (EURIPA 1981). Thus a recent report has noted "European information service providers have difficulties in achieving a big enough market, since the European market is effectively fragmented by numerous linguistic, technical and possible legal barriers. What actually exists is more a juxtaposition of national markets than a Community-wide market. As a result, it is difficult for the European information services to achieve financial stability, since investment in this activity is particularly risky and promotion of services costly" (Commission of the European Communities 1985, 7). Clearly therefore in sectors where economies of scale are important in information service provision, effective limits on the size of the potential market to European companies leads to a pronounced disadvantage compared with US competitors.

However, it should be acknowledged that data relating to, and information about, international information service transactions is extremely poor. As we have noted in Chapter 3 international service transactions cover a number of different forms associated with trade and foreign investment. Indeed, an increasingly important form of international transaction for information services is via communication networks. It should be recognised though that international information flows may be associated with the delivery of information from one part of a company to another or between different companies using the same data network and involve no financial transaction associated with the information interchange. With traded information services, on the other hand, the information is part of the product being sold as well as the service of delinking the information (Peat, Marwick, Mitchell & Co., 1986, 106). In this context it has been estimated that 70% of all International Information Flows (IIF) are intra-firm, in terms of being generated within trans-national corporations and flowing between headquarters and other branch units across the global (LOTIS, 1986, 18). More recently Sauvant has suggested an even higher figure noting that "perhaps 80-90% of trade (sic) in data services consists of non-commercial or intra-firm transactions "(Sauvant 1986c, 287). Moreover, most of these intra-firm IIF's involve only a handful of giant global corporations (OECD 1979). Even inter-firm traded international information flows are difficult to quantify, since there is no generally accepted method of valuing or counting information.

Clearly more research needs to be undertaken in traded information services and its impact on national and regional economies. However, given that the locus for information generation and delivery activities both within large corporations and in terms of the information producer companies (Chapter 5) is centred in the core regions of the Community it is likely that the less favoured regions of Europe will become net importers and consumers of information services rather than net exporters and producers. This aspect will be taken up in more detail in the next chapter dealing with the impact of the growth of communication networks on the development and location of information service activity.

Conclusions

This chapter has indicated the importance and growth of information services to the EC economy. In particular it noted the rapid growth of new and technologically advanced information sectors which not only had a direct economic impact in terms of employment and output expansion, but also indirectly via their impact on other sectors of the economy through their vital role as information and control channels. Key elements in the structure and functioning of these information service segments were then outlined. The importance of new and small companies in filling specialised market niches was noted within the overall context of increasing concentration of ownership in these information service markets. Large information service corporations, or 'majors', it was suggested benefitted from substantial economies of scale and high entry barriers and could be seen within a general context of sectoral and geographical diversification (on an international scale) combined with moves towards vertical integration. Many key entrants and subsequent actors in these information service markets originated from commodification and externalisation processes. However, it was noted that the EC was comparatively weakly represented in these new information service markets, with US companies playing a leading role in most EC countries and overall in terms of international trade flows and investment.

5 Networks, advanced telecommunications and the location of information services

The Role of Computer Networks and Advanced Telecommunications in the Development of Information Services

A key element in the growth and development of information services has been the introduction and spread of computer communication networks and developments in telecommunication systems. In terms of computer networks they are not only important in relation to intra-corporate information systems and how this may affect the organisation and location of corporate activity, but also in respect of how enterprises use networks to delivery and sell services to other companies. Their spread has been rapid. The number of transnational computer communication systems has, for example, doubled from 500 in 1979 to 1000 in 1982, with the number of terminals in Europe which can use such networks being forecast to increase from 625,000 in 1979 to 4 million by 1987 (Sauvant 1984).

Similarly developments in advanced telecommunication systems are having major implications for the growth and extension of the information marketplace and in terms of international information service transactions. As such there is increasing interdependence between the information services sectors and the communications industry on which they rely (Bruce, Cunard and Director 1986, 22). Thus information providers have established networks of transmission lines and switching capacity to offer a wide range of services to their customers on a regional, national and global basis. Customers will also utilise these networks to interact with vendors of services. Almost inevitably, therefore, the business of generating and selling information is becoming intertwined with the business of communicating over such networks. The importance of developments in computer networks and telecommunications is therefore associated with the wider issue of the role of communications in information generation and consumption

The vital role of communication in relation to the introduction of new and the development of existing information services is because information is a product which usually has to be consumed jointly with communications; only where the consumer produces the information himself is communication unnecessary (Flowerdew, Oldham and Whitehead 1983). Without communication information will often have little or no value. Much of the benefit therefore from information depends upon the ease of communication, its cost and the success with which information is communicated correctly and efficiently. As such developments in the means of communication and in their accuracy, and changes in the costs of different methods of communication have important implications for the effectiveness and development of information service provision and market growth.

The interface between telecommunication and information services can be mapped in a number of different ways, for example, in terms of demand and supply relationships or in relation to a spectrum of services ranging from :

(1) basic transport or bearer services;

(2) communication services;

(3) application services; and

(4) information services.

As with the use of the specific example of electronic information services (Chapter 4) there is increasing competition and merging of roles between the different types of communication and information service provision. There has therefore been increasing interaction between the various levels of supply as suppliers seek to protect their competitive position and increase the added value of their services in response to consumer pressure. As such a key problem is researching (and legislating for) these new information services is the fact that the definitions of different types of services are continually changing. The introduction of new technology in particular is transforming the capabilities of basic commucation transport services (through the introduction of stored programme switching exchanges and the emergence of ISDN and as such moving them into what is regarded as the realms of value added services). These technological advances are therefore progressively expanding the functionality of traditional carrier service offerings and leading inevitably to the blurring of the boundaries (Bleazard 1986, 7).

Indeed the impact of computer networks and telecommunications on the development of information services has been far reaching and all-embracing. Thus it has led to:

(1) the creation of new information sectors which has in turn increased the number and variety of information service markets;

(2) the geographical extension of the information marketplace;

(3) the increasing openess of national and regional information economies via increasing transborder information and data flows;

(4) this in turn is leading to fundamental shifts in the competitive nature and structure of the market in terms of the types of firms supplying the market and their location.

These processes outlined above will be taken up in the rest of the chapter.

Computer-Communication Networks and Information Service Location

Introduction: As noted earlier the spread of computer-communication networks is having a key enabling role in the introduction and growth of new information services (Hepworth 1986a; 1986b) have been important not only in how the internal information generation and processing activities are organised and located within companies but also in terms of how information services are generated, delivered and sold to other companies. It should be recognised though that intra-organisational private networks and delivery networks have become increasingly blurred with public telecommunication services as companies have extended their communication networks to cover inter-organisational linkages to connect with major customers, suppliers and financial institutions (Langdale 1983, 397) and which have then subsequently attracted other companies to share their network (Irwin and Merenda 1987).

The scale at which these networks operate range from a local or regional level to an inter-national scale, associated with developments in Local Area Networks (LANS) and Wide Area Networks (WANS). Aside from the spatial structure of networks more specific operational details of network configuration relate to levels of service provision and the organisation and complexity of switching centres, computer, capabilities and user terminals (Dunn 1977, 109-10). Much of the discussion in relation to the development of computer-communication networks and its impact on service location and regional development has been covered in the UK (Howells and Green 1986a, 162-189; 1988). Some of the key elements in the discussion from that report are outlined here.

Networks and Corporate Organisation: The spatial configuration of a corporate computer-communication system not only reflects the geographical organisation of a company in terms of the locus of control and decision-making, its information flows and its functional layout, but also once established it can in turn influence corporate organisation. The introduction of private or leased-line networks arose with the growth and development of large multi-site companies which found that a significant part of their overall communication costs were associated with intra-organisational communication linkages between different sites (Langdale 1982). Companies with a high volume of intra-organisational information and data traffic between a large number of units can therefore make substantial cost savings when deciding to operate a private service network. However, efficiency and control factors would also appear to be additional incentives to operate private communication networks. Such networks improve the flow of information between an organisation's functions and units allowing more effective decision-making, better communication and feedback of ideas.

Most computer communication networks in companies frequently evolve out of existing communication links. Thus Read (1977, 128) notes that most networks that have been established by companies grew in a haphazard way and were overall hybrids of former voice, data and computer communication links. These former systems have subsequently been acquired and integrated into a service system in response to newly perceived management information requirements, available technologies and cost efficiencies. Once established however networks can help to influence the spatial development and layout of a companies' functional organisation and structure.

In the context of specific case studies and examples of computer-communication networks and in relation to corporations operating in Europe, Bakis (1980; see also Bakis 1982; 1985) has provided a detailed analysis of IBM's communications network in France and on an international basis. In the context of IBM-France's communication network within France, Bakis notes that it is a mixture of a private regional network and long-distance public switched network.

The private network consists of six main switching centres located in the Paris region which is highly interconnected carries high traffic volumes and associated with this links information and data intensive functions. IBM-France uses, for cost reasons, a public switched network for longer distance contacts with factories and offices which have much lower information needs and which therefore generate much lower traffic demands (Bakis 1980, 26-31 and 51). An exception to this is the Sainte-Marie in Orleans which is linked into the private network system of the company because of its high traffic volumes with other IBM centres in Paris and therefore can be seen as part of the organisation's Paris information system.

Bakis (1980, 44-53) also goes on to outline the framework of IBM's global network. The main national concentrators (computer centres linked to the IBM network and to network access points) in Europe are all hubbed around the Cosham centre near Southampton in the UK (part of the 'Respond' system under the control of IBM-Europe) which in turn is connected to the IBM-USA organisation via the White Plains (New York) centre and IBM World Trade Corporation's (IBM-WTC) America and Far East network. Thus, for example, the IBM-United Kingdom centre is linked into the main 'Respond' centre at Cosham. The wider implication of the development of this global network by IBM is that it has allowed it to develop a high degree of production specialisation between its individual European plants (making nationalisation extremely difficult) but has also allowed to have a highly cohesive and integrated international manufacturing system.

Corporate networks are also being established on a wider international bases, particularly in relation to the financial services market. Lloyds Bank International (LBI) is setting up a communications network which will link up its visits in 47 countries across the world, using the Phillips Sopho-net system. The LBI network handles all data and text traffic including letters of credit, electronic mail and foreign exchange deals and trunk voice traffic as well. The system replaces a telegraph message switching network which is still be used for funds transfer and in turn has been incorporated into the new network.   The introduction of the network is being phased over four years.   The first phase which is now completed is shown in Figure 5.1 and began by linking three major network nodes, two in London and one in New York, which formed the main data switching and conversion centres. The development of this new communication network by LBI is in response to its own corporate requirements in terms of information and data handling and in the increased global requirements of large corporate customers which want improved financial information from their bank in the different countries in which they operate.

The development of private computer communication network systems in large corporations can therefore be seen as an evolutionary process, building on pre-existing systems and outlining the basic organisational framework of the company.   Their establishment has arisen out of a response to reduce intra-corporate communication costs and improve the efficiency of internal information and data flows, in turn fulfilling other management criteria and which has been associated with the increasing specialisation of informatics workers (Perring 1983, 101).

Such systems are also often subsequently expanded to fulfil the requirements of a company's external or inter-corporate linkages with its major customers and suppliers by linking them onto the network.   On a more specific spatial basis, networks not only serve to reinforce the corporate hierarchy of decision making and control but also reflect the development of a company's organisational structure in response to growth and changing market conditions.

Before moving on to discuss delivery networks it is important to acknowledge the major input that computer-communication networks are having on the organisation of location of information service activities within large corporations and its consequent job and economic growth prospects (Hepworth 1986a, 417).   Some decentralising tendencies have been revealed with, for example, data entry and certain kinds of printing and processing functions being decentralised to agency or 'field' offices (Baran 1985, 93-4).   Certainly in relation to integrated production and service facilities these have become locationally more

FIGURE 5.1

LLOYDS BANK INTERNATIONAL WORLDWIDE DATA NETWORK : PHASE 1. 1985

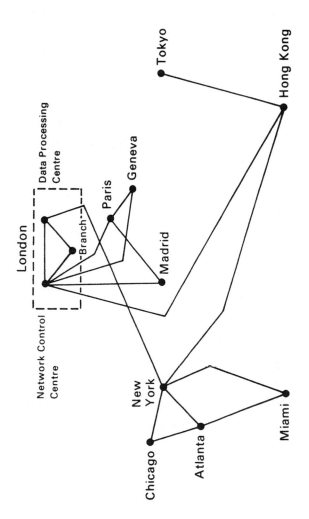

Source : Howells and Green 1986a, 170

137

footloose in terms of proximity of their customer base allowing them to relocate in less expensive, decentralised sites but often this has been restricted to a few localities where in reality good quality clerical labour and telecommunication 'hookups' are available. Overall, however, although networks allow a more decentralised system of strategic functions and operation it has been found that companies are using their systems to maintain and support centralised organisational structures (Bakis 1980, 87). Thus there has been a strong centralising tendency with the introduction of new computer and telecommunication networks which has led to spatial consolidation and closure of local and regional offices. Obviously however, these pressures for centralisation and decentralisation are highly inter-related and complementary, with data entry being dispersed whilst direction, control and decision-making being highly centralised.

Delivery Networks and Markets: Computer communication networks, however, can also be used to connect the needs and resources of users to the capabilities and services of information producers and facilitate transactions between them (Dorkick et al. 1979, 220). A whole range of information and data transactions can be provided on such an electronic information (Thompson 1971) or network marketplace (Dordick et al. 1981). Four main types of delivery networks have been suggested (see Todd and Strasser 1985, 11):

(1) Proprietory networks (for example, Citinet, Telerate, Reuters);

(2) Third-Party networks (for example, Geisco, Istel, I.P. Sharp);

(3) 'Mutual' or 'League' delivery (for example, SWIFT; EURONET; SITA; and WMO/GTS); and

(4) Public packet and circuit-switched and satellite networks.

In relation to (3) SWIFT, as an international 'mutual' network was initially set up in 1973 by a number of European banks to act as a cheap and secure message switching facility to handle international transactions covering bank transfers, foreign exchange, confirmations, documentary credits and query messages (OECD 1979, 201). Its network coverage has been gradually extended to cover Australia and New Zealand in 1982 and South East and East Asia in 1985 (Langdale 1985 , 7), although Europe still dominates the geographical distribution of traffic flows. By 1985 some 1257 banks in 54 countries were using the SWIFT network, which is centred on La Hulpe in Belgium. However, although SWIFT is being upgraded as part of the SWIFT II programme, its position as the banks' primary telecommunication network is being eroded as banks use their own networks to communicate with their branches. Another example of a international mutual network is SITA, the international airline reservation network, and EURONET, the European Community's public data network. Two examples of national scale mutual of league delivery network systems operating in the UK are MATRIX and LINK. MATRIX represents a cash network which is operated by seven large UK building societies (home loan organisations) : Anglia, Bradford and Bingley, Leeds Permanent, the Alliance and Leicester, National and Provincial, Woolwich Equitable and the Bristol and West. The shared network links in the cash dispensers and ATMs of these building societies. By contrast LINK represents a larger and more diverse group of 21 banks and building societies operating a shared network of ATMs in Britain. A further UK example is the London Stock Exchange's computer network system, TOPIC, which covers key financial centres in Britain and Ireland, as well as the offshore centres of Jersey and the Isle of Man (Hepworth 1987, forthcoming). However, approximately 80% of all dedicated terminals used by TOPIC subscribers are located in the Greater London area, again emphasising the dominance of London as the financial and information service centre of the UK.

In relation to the proprietory and third-party delivery networks in Europe many have arisen out of activities which served a company's own internal requirements. Thus GSI Alcatel the computer software company has developed one of the largest computer-communication networks in Europe (Figure 5.2). This has stemmed out of its own internal corporate information network which has been improved and developed so that clients can have direct access to its processing potential. The network is hubbed around two international 'production' centres in Paris and Grenoble and two satellite centres in Geneva and in Darmstadt. In 1985

FIGURE 5.2

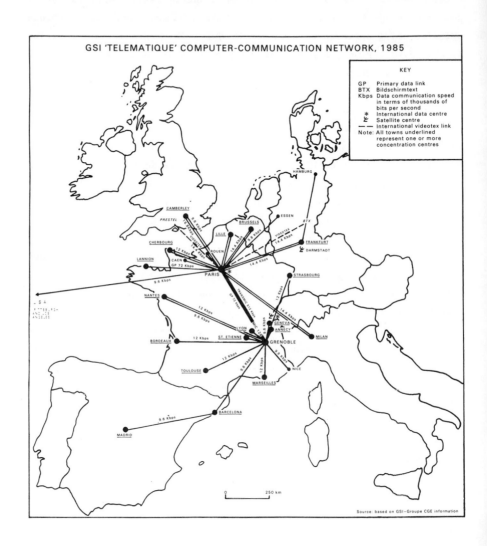

GSI 'TELEMATIQUE' COMPUTER-COMMUNICATION NETWORK, 1985

KEY

GP   Primary data link
BTX   Bildschirmtext
Kbps   Data communication speed in terms of thousands of bits per second
   ✳   International data centre
   ⚸   Satellite centre
— —   International videotex link
Note: All towns underlined represent one or more concentration centres

Source: based on GSI—Groupe CGE information

GSI linked its own computer-communication network with public networks, such as Transpac, Tymnet and PSS, and also via public videotex systems in France (Teletel), UK (Prestel) and Germany (BTX). Thus GSI runs an airline reservation service in France (Telelut) and Germany (Maris) using the public Teletel and BTX videotex network which allows airlines, vehicle rental agencies and hotels to communicate with travel agents and tour operators. Similarly in Britain the largest privately self-managed integrated network is run by ICL which uses BT leased lines and which offers a wide range of services to third parties.

On an international level Reuters is a prime example of a major network delivery operator, providing a range of information and data communication services ranging from media and financial data to quotation retrieval, dealing communications and dealer interfaces. Reuters developed its operations on an extensive international leased networks and has been able to 'piggyback' additional business and financial services onto its basic network. The Reuter communications network now has 276 international circuits connecting 118 countries, with high speed (56,000 bits per second) being introduced on its major routes. The network comprises many different communications technologies including point-to-point satellite and small dish systems, micro-wave and co-axial cable systems as well as traditional leased telephone lines. Recently, Reuters has expanded its operations by acquiring the international new pictures business from United Press International (UPI). Indeed UPI has its own extensive private line network with switching centres based in Brussels, Hong Kong and New York (Read 1977, 128).

Telecommunications and the Development of Information Services

A key element in the development and location of information services is the structure and spatial availability of public telecommunication networks. The spread and development of information services, which are transaction orientated by nature, are heavily dependent on both national and international telecommunication networks (Bruce, Cunard and Director 1986, 15). However, Gillespie et al. (1984, 61) have indicated that there are significant national and regional disparities within Europe in terms of the supply of telecommunications infrastructure and that there are little signs of convergence.

In particular Gillespie et al. (1984) have shown there are substantial regional variations in the number of telephone (Table 5.1) and telex (Table 5.2) subscribers and in the concentration of national network terminating points (Table 5.3). The study by EIU/CURDS (1985; see Howells and Green 1986a, 176-9) has indicated the way in which regional disadvantages in the telecommunications supply, based on the Northern region of the UK, can affect the provision and growth of information services in such localities. Thus the research found that for basic telecommunication services the main disadvantage for the peripheral Northern region was largely one of cost, although for more advanced switched services there were more serious deficiences, namely lack of provision in many areas (Table 5.4). As such access points to specialist (mainly packet switched) data-communications networks in the less favoured regions of the Community are only available in the larger metropolitan cities (Pye and Lauder 1987, 101). Clearly the provision of basic and advanced telecommunication services and computer-communication networks can have a profound influence on the spatial growth and development of information service activities.

In turn a key element in the provision and costing of telecommunications services has been the regulatory and public policy environment. Thus "in terms of internal variations in levels of service and tariff levels within countries there can be little doubt that public sector monopoly provision in Europe has benefited peripheral and/or less favoured regions" (Gillespie and Hepworth 1986, 23). It is also associated with other policy factors such as in Canada where it has been stressed that the reduction inthe regional variation in household telephone coverage and the eventual universal household was due in part to a policy commitment on universality of access (Pike and Mosco 1986, 22). However, with the gradual shift towards deregulation and privatisation as has occurred in Europe (Bruce, Cunard and Director 1986, 130) the less favoured regions of the Community are now coming

TABLE 5.1

EEC NATIONS : SUMMARY OF REGIONAL VARIATIONS IN THE NUMBER OF TELEPHONE SUBSCRIBERS PER 100 INHABITANTS, 1980-1

Nation (number of regions)		Region	Telephone subscribers per 100 inhabitants	As % of national average
United Kingdom (10)	highest:	London	42	127
	national average:		33	
	lowest:	Northern Ireland	23	70
Germany (10)	highest:	West Berlin	53	156
	national average:		34	
	lowest:	Regensburg	24	71
France (21)	highest:	Paris	40	133
	national average:		30	
	lowest:	Franche Comte) Lorraine)	23	77
Italy (20)	highest:	Liguria	36	156
	national average:		23	
	lowest:	Calabria	12	52
Netherlands (13)	highest:	Amsterdam	43	126
	national average:		34	
	lowest:	Hengelo	30	88
Denmark (3)	highest:	Sealand/Mon	49	109
	national average:		45	
	lowest:	Mid and N Jutland	41	91
Belgium	highest:	Brussels	36	133
	national average:		27	
	lowest:	Hasselt	19	70
Greece	highest:	Athens Region	35	141
	national average:		26	
	lowest:	Thrace	9	38
Ireland	national average:		21	
Luxembourg	national average:		36	

Source: National PTTs; from Gillespie et al. (1984, 51).

143

TABLE 5.2

EEC NATIONS : SUMMARY OF REGIONAL VARIATIONS IN THE NUMBER OF TELEX SUBSCRIBERS PER 100 INHABITANTS, 1980-1

Nation (number of regions)		Region	Telephone subscribers per 100 inhabitants	As % of national average
United Kingdom	highest:	London	0.41	256
	national average:		0.16	
	lowest:	Wales	0.08	50
Germany	highest:	Hamburg	0.43	146
	national average:		0.23	
	lowest:	Kiel	0.12	48
Netherlands	highest:	Amsterdam	0.47	204
	national average:		0.23	
	lowest:	Leeuwarden	0.10	57
France	highest:	Ile de France	0.30	188
	national average:		0.16	
	lowest:	Bas Normandie	0.08	50
Italy[1]	highest:	Lombardia	0.14	175
	national average:		0.08	
	lowest:	Basilicata	0.01	13
Greece	highest:	Athens	0.25	166
	national average:		0.15	
	lowest:	Thrace	0.05	33
Belgium	highest:	Brussels	0.56	254
	national average:		0.22	
	lowest:	Libramont	0.06	27
Denmark	national average:		0.20	
Luxembourg[1]	national average:		0.52	
Ireland	national average:		0.15	

[1] 1982 figures

Source: National PTTs; from Gillespie et al. (1984, 51).

TABLE 5.3

SHARE OF NATIONAL NETWORK TERMINATING POINTS ACCOUNTED FOR
BY THE LARGEST REGION, 1979

NATION	HIGHEST SHARE REGION	SHARE OF NATIONAL TOTAL
France	Paris	43%
Belgium	Brussels	48%
Denmark	Copenhagen	37%
Germany	Dusseldorf	10%
Ireland	Dublin	79%
UK	London	43%
Netherlands	Amsterdam	25%
Italy	Lombardia	30%
Greece	Attika	93%

Source: Eurodata Foundation Reports – "Data
Communication in Western Europe in the
1980s"; from Gillespie et al. (1984, 54).

145

TABLE 5.4

UK NORTHERN REGION DISADVANTAGES IN TELECOMMUNICATIONS SERVICE SUPPLY

TYPE	SWITCHED	LEASED
Traditional	Relatively small local calling areas Fewer low cost routes	Expensive to install Network, eg. from SE to Northern Region
Data	Higher costs for occasional access	Expensive to install Network, eg. from SE to Northern Region
Mobile	Available only around trunk roads	-
VANS	Higher costs for occasional users and Prestel users	-
Digital	No IDA trial	As above. BT services available only in main towns. Mercury not available at all

Source : EIU/CURDS (1985, 143).

under threat. This shift in terms of deregulation and privatisation has been most marked in the UK where the provision of new telecommunications services in the peripheral, less favoured areas of the country looks increasingly in doubt (Howells and Green 1986a). Certainly more research needs to be undertaken on the regional impact of deregulation and the more market driven approach towards telecommunications provision.

The regulatory and wider policy framework of telecommunication services clearly has an impact on the growth and development of new and advanced information services in the EC. On one hand it can be argued that deregulation and privatisation int he UK has helped to develop the overall growth of communication and information activity at a national level. Thus associated with shifts in the regulatory environment, increased competition and relatively low international telecommunication changes (in conjunction with importance of the City of London in the international services field), has encouraged US information service majors to centre their European operations in London (Information Dynamics Ltd 1986, 74) and communication and related service activity has grown rapidly in the UK. Thus it has been estimated that the UK has over a third of all the European market in electronic information in 1984 (Table 5.5). Similarly the UK has the largest VANS market in Western Europe (Frost and Sullivan 1985) with over 800 VANS licenses granted in early 1987 (Howells and Green 1988).

By contrast, the French government with a strong public orientated telecommunications regime has also been successful in the development and provision of new forms of telecommunication services. Thus the DGT has developed a network of terminals for an electronic telephone directory based on a videotex system, Teletel (using Minitel), to households in France. It has been estimated that were at the end of 1986 over 3 million Minitel terminals installed across France, establishing the world's largest and most successful operational videotex system.

TABLE 5.5

NATIONAL SHARES BY SALES TURNOVER
OF THE EUROPEAN ELECTRONIC
INFORMATION INDUSTRY, 1984

COUNTRY	% TURNOVER
UK	34
Germany	21
France	19
Italy	6
Benelux	4
Denmark	3
Greece	1
Ireland	<1
Switzerland	3
Austria	3
Scandinavia	8

Source : 'Survey of the European
Information Industry :
Its Electronic Developments'
European Information Providers Association
(EURIPA) 1985.

Networks, Telecommunications and the Location of Information Services

As we have seen therefore computer and telecommunication networks are important in relation to the spatial growth and development of information service activity both on an intra-corporate and an inter-corporate, delivery market basis. Computer networks have been hubbed or centred around key strategic and data processing units and reflect the volumes and types of information and data communication that a company transmits. On a national level communication centres and data processing facilities have been located near to company head offices located in the main metropolitan centres of national territories.

On an international level key locations are emerging as sites for data and communication networks (Moss 1986a; 1986b; 1986c). New York, London and Tokyo have benefited from being key international financial centres and the hubbing of communication networks around these centres has tended to reinforce their position. Nevertheless large multinational companies are to some extent becoming more flexible and mobile with regard to the location of information and telecommunication intensive functions. Tariff changes, regulatory environment and service availability as we have noted are important when considering information and communication centres on an international level, in addition obviously to the company's existing corporate structure and location.

This is evident in the UK which has benefited in the past from tariff changes and regulatory attitudes (Howells and Green 1986a, 174-5). Thus when the Bank of America was rationalising its European communications network, via the closure of one of its two major communication centres it decided to close down its centre in Frankfurt and concentrate its communication activities in London. The US Chase Manhattan bank also has its European data centre located in Britain at Croydon together with IBM's major communication centre at Cosham, whilst Dun & Bradstreet's European data base is located at Harefield near London. All these shifts and developments in the corporate geography of networks therefore have a key role in the focus of information gathering and distribution and decision-making on a national and international scale. A key location factor for US companies locating in the UK is however also the importance of London as a key global financial centre, and on-line financial services, as noted earlier, are an important 'lead' electronic information service sector. Thus the UK is estimated to have 350,000 Network Termination Points, some 25% of the European total, of which 28% are in the financial sector by far the dominant sector (Information Dynamics Ltd. 1986, 71). However, undoubtedly it is a reflection of British Telecom's pricing structure in the past in the context of international calls, particularly to the US. This is evident in a recent survey (Table 5.6) which indicates the savings that could be made re-routing calls to the US using British Telecom transit switching services. Only in Italy, France and more particularly Sweden were no savings made re-routing calls via the UK in this way.

However, other EC countries are keenly aware that it is important for them to attract large multinationals to 'hub' their European networks in their countries. Indeed in a survey of comparative telecommunication volume and charges (for a user making a search on a free database) in 1987 the UK was now ranked 8th in terms of cheapness both for national and European telecommunication links (Table 5.7). The most expensive countries in the Community in terms of telecommunication changes were Germany, Spain and Italy (for European calls). The cheapest by contrast were Portugal, Luxembourg and above all Ireland. Another dimension not revealed in this table is the call failure note on public data networks where there appears substantial variation in terms of failure rate between member states in Europe (Nairn 1987, 46), due primarily to a lack of standardisation of prompt and network error messages. Not surprisingly Ireland aims to become a major information and data processing centre in Europe (Chapter 6) based on its improved telecommunications network and now having one of the lowest telecommunication changes in Europe (Price 1987). By contrast Belgium has certain locational advantages in the sense that it is essentially the headquarters for the Commission for the European Commission, is the headquarters site for SWIFT, the international bank transaction league network and it is also the European headquarters for many US companies such as 3M, Proctor and Gamble and ITT. In the Netherlands, Rotterdam

TABLE 5.6

<u>INTERNATIONAL TELEPHONE COSTS FROM EUROPE TO THE USA :</u>
<u>POTENTIAL SAVINGS USING BRITISH TELECOM TRANSIT SWITCHING SERVICES</u>

	TELEPHONE COSTS TO USA DIRECT (ECUs)	TRANSITING COSTS VIA UK	NET SAVINGS +/-
UK	1.25	–	–
Sweden	1.47	2.08	+ 41.8
Denmark	1.86	1.71	– 8.0
France	1.89	1.91	+ 1.2
Netherlands	1.90	1.70	– 10.7
Luxembourg	1.96	1.69	– 13.4
West Germany	2.11	1.85	– 12.4
Italy	2.30	2.35	+ 2.3
Greece	2.30	2.24	– 2.8
Ireland	2.34	2.07	– 11.4
Spain	2.97	2.42	– 28.5
Belgium	3.17	1.91	– 39.6

<u>Source</u> : based on Information Dynamics Ltd. (1986, 128; using data
from Director des Telecommunications des Reseaud Enterviews
of Direction Generale des Telecommunications, France).

TABLE 5.7

COMPARATIVE TELECOMMUNICATIONS VOLUME AND TIME
CHARGES IN EUROPE 1987*

	National	Rank	European	Rank	Barrier Factor: % European National	VAT
		(1 =	Cheapest)			
Austria	3.56	15	4.987	17	140%	0%
Belgium	1.596	6	3.171	4	199%	19%
Denmark	1.504	5	3.982	7	265%	22%
FR Germany[2]	4.855	16	5.754	13	118%	0%
Finland	1.965	11	5.808	14	296%	16%
France	1.914	10	4.637	12	242%	19%
Greece	Not available		6.567	15	Not available	12%
Ireland	0.90	1	2.252	1	250%	0%
Italy	1.847	7	4.643	16	251%	18%
Luxembourg	1.25	3	2.622	2	210%	12%
Netherlands	2.031	12	3.517	5	173%	0%
Norway	1.906	9	3.803	6	203%	20%
Portugal	1.162	2	3.030	3	261%	0%
Spain	2.487	13	4.238	11	170%	12%
Sweden	1.541	4	4.180	10	271%	0%
Switzerland	2.500	14	4.080	9	163%	0%
United Kingdom	1.905	8	4.024	8	190%	15%

*   For a user making a search (20 minutes searing at 1200 band and using 620 segments of digital information; 1 segment = 64 bytes of characters) on a free database. The user is dialling up to a PAD and paying a local telephone call.

1   The barrier factor gives the increase the user is likely to pay as soon as he/she transmits data across the national border.

2   The German figures use the increased telephone charge from 1 April 1987 which make a substantial part of the costs of communicating. The access charge amounts to about half the total cost.

Note: In the table no account has been taken fixed costs like subscription to the network or hire of PTT equipment. Nor as noted above are database search costs included. The payment in local currency has been converted into ECU at the current rate for 5 February 1987. The costs are based on the latest 1987 tariffs released by the PTTs. Dial up charges to the node are included in these figures. VAT has been added where the PTT tables have indicated it is extra: otherwise it is assumed that VAT is included in the published charges.

Source: adapted and compiled from Information Market 47, 8–9; using data from Directorate General XIII, Commission of the European Communities, Brussels.

is being developed as a European 'hub' for electronic <u>information</u> and data interchange in order to safeguard the international competitive position of Rotterdam as the world's busiest <u>bulk-goods</u> port (Netherlands Economic Institute 1986, 67-72). The 'Teleport' scheme is designed to help ease Rotterdam's paperwork congestion. Two schemes : International Transport Information Systems (INTIS) and Cargonaut, centred at Schiphol Airport, associated with this strategy and are systems for integral control of goods and information flows. Similarly Transpatel, an open information system used for freight allocation, based on the Dutch PTT videotex service is also being linked with five other European videotex systems (Netherlands, UK, Federal Republic of Germany, Switzerland and Austria). Finally, in France the city of Metz is also building a 'teleport' operations to offer discounts for international communications.

Conclusions : Communications and Information Service Location

Key elements in the location of newly emerging information services were outlined including amongst others the development of corporate computer-communication networks, the growth of major international information gatherers and distributers and the spatial provision and pricing of telecommunication network services, associated with the regulatory environment of communication services overall. Certain specific centres were seen as developing as information complexes both on a national and international basis, centred on the core regions of EC national territories. Increasingly in Europe 'information rich' and 'informtion poor' regions are developing. For example, in the UK there is an increasing dichotomy between the South East with London being a key international information generating and processing centre and market, and 'information poor' regions, such as the North, with their low level, routine information and decision-making activities. As such strategies need to be developed to unlock the less favoured regions of the Community from the long-run cumulative processes which influence the agglomeration of key information services.

6 Services in regions and the economic development of the European community

Introduction

This chapter will attempt to focus on the overall policy framework and key issues relating to the regional economic growth and service development in Europe. The long term policy issues raised here and elsewhere (Howells 1987a) can be viewed within the wider framework of the 'FAST, 1984-1987 Objectives and Work Programme' (Commission of the European Communities 1984a), in particular associated the production of a strategic dossier by the Commission 'Services, Infrastructure and Regions', which in turn will be incorporated in a Proposed Initiative by the Community (PIC) under the title of 'The Community and Services'.

The rest of the chapter will examine the role of services in regional economic development and then briefly review those policy areas within the remit of various Directorate-Generals (DGs) which cover services. The last section will present briefly some of the key

strategic issues which are seen as being important in the creation of a
policy framework for encouraging service development in the regions.

The Role of Services in Regional Economic Development

Services and Economic Development: Traditionally services have been
viewed as 'passive' elements within modern industrial economics with the
growth of services in regions simply following that of population and
economic growth. There is, however, increasing evidence to suggest that
this is no longer an adequate conceptualisation of the place of the
service sector in national and regional economic development. Indeed
research suggests that in some respects the service sector is the
leading rather than a lagging element in national and regional economic
growth in many parts of the Community.

However, it would be misleading to suggest that all service sectors
play a vital role in economic growth - there are important distinctions
to be made. A key difference to be made is between basic (exportable)
and non-basic or induced (locally oriented) services. It has already
been noted that the traditional view of service activity was that it
passively followed socio-economic change in the rest of society. This
was because services were largely seen as non-basic, induced activities
which served purely local consumer needs, had limited multiplier effects
and which simply reacted to change. By contrast basic service
activities have largely been ignored by researchers. These latter
activities can be defined as those services which are geared for
national and international markets and often provide a substantial net
balance of payments contribution to an area or country. In addition
such services are largely able to maintain self-sustaining growth
independent of a particular locality and provide significant multiplier
effects on a local, regional and national basis. Information and
producer services such as banking, financial and computer services are
obvious examples of service sectors which have a significant basic
component.

Producer Services and Economic Development: A key category of services which are associated with an orientation towards basic (exportable) services are producer (or intermediate) services which can be defined as services which provide output which is consumed or used by firms, for example market research, accountancy, management consultancy and financial services. The key role that producer services and business play in modern industrial economies (Marshall 1985; Wood 1986; Krollis 1986) and therefore their policy importance is strengthened by the fact that producer services are important sectors for output and employment growth in the economies of the Community.

Other service sectors, or service functions within firms, however, can also be important in regional policy terms. Tourism is one such sector which has an important tradeable and growth role in many less favoured regions in southern Europe. Similarly headquarters and R & D functions also have a significant impact on their local and regional economies. Thus headquarters of large corporations not only provide a source for high level, strategic service functions and therefore jobs but a high proportion of key external information and business service requirements are sourced within the local area of the headquarters site. Similarly R & D establishments of large corporations located in less favoured regions can have a key role to play in the development of research and technology, and hence economic growth of such localities.

Information Services - Key Change Elements in the Economic Development of the Community: By concentrating on producer service activities as a policy for services in the European Community there will be a tendency to ignore the profound changes that are occurring in the European economy associated with the growth and development a wider and more pervasive service group, information services. As such a purely producer service policy focus will also be in danger of taking a narrow sectoral view of service policy. It will tend to neglect the broader economic role of services and information in other sectors of the economy. Related to this it will tend to departmentalise producer services in their own narrow sub-sectors, such as market research and accountancy, instead of acknowledging the growing inter-links and convergence of services within the wider framework of the evolution of 'meta' industries.

The orientation of a service policy framework on information services, it is presented here, would also more firmly integrate with the policy objectives of other Commission DGs in terms of services. Thus, for example, information services are seen as central elements in the international trade sphere (DGI), the creation of the internal market (DGIII) and above all in the development of information and telecommunication industries in the Community (DGXIII).

The reasons why there should be a focus on information services for a services strategy for the Community has been outlined earlier (Chapter 4). These include the key role that information plays in the increasingly integrated economic system of industrial economies; the rapid and substantial growth of information services in Europe; the pivotal and integral role that information services play in the technological and structural trends in the European economy; and the important function that information services play in the growth and development of firms.

Services : The Policy Context within the European Community

Introduction: A key element in the development of a strategy or set of strategies towards services in the context of regional economic development in the Community is recognition of the existing policy framework. The following sub-sections therefore outline policy fields and issues which touch upon services amongst the relevant DGs in the European Commission. In total nine DGs have been judged to have at least some relevance to the development of service activity in the Community (Table 6.1).

(1) External Relations and Services: The EC has the responsibility for negotiating on behalf of the Member States discussions relating to trade under the General Agreement on Tariffs and Trade (GATT). "The services sector in general, and tradeable services in particular, are of

TABLE 6.1

EUROPEAN COMMISSION DIRECTORATE GENERALS' HAVING RESPONSIBILITY
FOR POLICIES WHICH RELATE TO SERVICES

Director General	Title and Coverage Relating to Services	Sub-Section number (where policy context described)
DGI	External Relations (including GATT negotiations)	1
DGII	Economic and Financial Affairs (including economic integration and sectoral i.e., service issues)	2
DGIII	Internal Market and Industrial Affairs (including removal of internal barriers to trade, freedom of establishment and freedom to provide services)	3
DGIV	Competition (including international aspects, restrictive practices in services, joint ventures and mergers).	4
DGV	Employment, Social Affairs and Education (including employment policy)	5
DGVI	Agriculture	6
DGXII	Science, Research and Development (including FAST and CREST)	7
DGXIII	Telecommunications, Information Industries and Innovation (including telematics, new technologies and information networks together with the Information Technologies Taskforce which covers ESPRIT, etc).	8
DGXVI	Regional Policy (including regional policy for services)	9

major importance to the Community economy, particularly as a determinant of industrial efficiency and as a source of new employment opportunities and of balance of payments earnings. The potential of the sector to increase still further these contributions should be actively explored" (Second Report of the Inter-Service Group on International Trade in Services, EC). Currently there is a new round of GATT negotiations of which a major element is that of services, and more particularly whether trade in services should or can be legally organised under GATT.

The Commission, via DGI for External Relations, in light of the discussions on international trade in services going on in the Inter-Service Group, has set up a Work Programme which it can use on a day-to-day basis to reach decisions and achieve long-term aims for the Community. For example it has resolved that discussions on international trade in services has to include debate about the right to, and conditions attached to, 'establishment' (i.e., foreign investment in the form of offices or subsidiaries located overseas). It has also concluded that since considerable differences exist between types of service sector the same solutions may not be appropriate to all of them. This suggests that a sector-by-sector approach for services should be pursued.

Discussions about services in relation to GATT negotiations are continuing and it is obviously impossible to predict what the final outcome of the discussions will be. However, the US, the European Community and to a lesser extent, Japan, are all keen to include services under the current round of GATT negotiations. As such extra-Community trade and 'establishment' in services is likely to become more open over time.

(2)   Services and the EC Economy: Obviously services form a key element in the overall economic structure of the Community, and as such it represents a key monitoring area for Community intervention. In this respect Directorate - General II for Economic and Financial Affairs has clear involvement with services. This has to do not only with issues relating to economic structure and problems of economic integration, but

also with more specific sectoral and structural difficulties associated with services.

(3) The Internal Market for Services and Less Favoured Regions in the
Europen Community: In the Commission (via DG III, the Directorate-
General for Internal Market and Industrial Affairs) the establishment of
a Common Market in services is seen "as one of the main preconditions
for a return to economic prosperity" (Commission of the European
Communities 1985a, paragraph 95). However the freedom to provide
services across internal frontiers has been much slower than the
progress achieved on the free movement of goods. As such, the Commission
considers that swift action should be taken to open up the whole market
for services. This applies both to the new service activities, such as
information marketing and telematics, and as well as to more traditional
services, such as transport and banking, which can play a key
opportunity supporting role for industry and commerce.

Service sectors issues specifically considered in 'Completing the
Internal Market' paper are:

(1) financial services, (including banking, insurance and securities);
(2) transport;
(3) new technologies and services; and
(4) the information service market.

Some of the key issues raised in the White Paper are:

(a) Harmonisation of supervision of financial services in the
national markets of the EC should by the principle of "home
country control". This means attributing the primary task of
supervising the financial institution to the competent
authorities of its Member State of origin.

(b) Decision to create a European securities market system, based
on Community stock exchanges which are linked electronically.
The action is designed to break down barriers between stock

exchanges and to create a Community-wide trading system for international securities.

(c) The development of new technologies has led to the creation and development of new cross-border services which are playing an increasingly important role in the economy. However, these services can only develop their full potential when they serve a large, unobstructed market. This in turn necessitates the installation of appropriate telecommunication networks with common standards.

(d) The information market in the EC is undergoing far-reaching changes associated with new information technologies. Such changes are associated with:

- the exponential growth of information;

- the growing speed with which new information becomes obsolete;

- the strong tendency of information to flow across borders; and

- the application of new information technologies.

Information itself and information services are becoming more and more widely traded and valuable commodities, and in many respects represent primary resources for industry. The opening of the market is therefore becoming increasingly essential. Moreover the functioning of markets for other commodities depends upon the transmission and availability of information.

(4) Competition Policy and Services: DGIV, the Directorate-General for Competition is involved with international aspects associated with competition, the problem of restrictive practices and issues relating to joint ventures, merger activities and the role of public enterprises, state monopolies and public discrimination. Although services have traditionally not had a central role of DGIV, it does have a specific

division under Directorates on Restrictive Practices and Abuse of Dominant Positions in relation to Distribution, Banking and Insurance, the Media and Other Service Industries, whilst under Directorate D it has a division responsible for Industrial and Intellectual Property Rights, a increasingly important issue in relation to information services. In addition DGIII has undertaken studies into the international competitiveness of the European Communities tradeable service sector.

(5) Employment and Services: DGV for Employment, Social Affairs and Education is obviously clearly involved in the service sector under its employment remit, since this sector from the mid 1970's onwards has been the major provider and growth sector for jobs in the Community. Thus in a report for DGV it was noted that "The service sector was the main source of newly created jobs during the seventies. Indeed service employment actually exceeds employment in industry in all Member States. It is the only major sector in which employment continued to grow during the seventies both in terms of numbers and shares of total employment" (ST 81/29, 9).

(6) Services and Agriculture: Within agriculture the conjunction of trends in consumer behaviour, the food industry and the policy paradigm are creating opportunities in which new policy directions can be pursued to achieve stated policy objectives. Policies under the remit of DGVI for Agriculture applicable to services are associated with its remit to support the development of self-sustaining, economically viable rural communities and obviously services have an important role to play here. For example in terms of the role that financial and business services may have in the formulation of farmers' co-operatives and joint ventures associated with food manufacture; the need for improved marketing services for agriculture; the role that research services have to play into developing new market niches for food; and finally, the role of communication and distribution services in improving information and commodity flows between and within rural areas.

162

(7) <u>The Transformation of Services and Technological Change in the Community</u>: Directorate General XII for **Science, Research and Development** are directly involved in the prospects of services development in the EC in association with technological innovation via the FAST programme. Research activities undertaken in FAST I programme indicated the challenges posed to the European Community by the development of service activities, particularly associated with technological change. Arising from the research programme DGXII aim to produce a Proposed Initiative by the Community (PIC) designed to encourage the integration of the various aspects of Community policy relating to services and identifying specific priority initiatives to be undertaken jointly in order to provide a long term response to the structural and territorial challenges presented by the growth and development of services in the Community.

(8) <u>Telecommunications, Information Industries, Innovation and the Regions</u>: The newly enlarged DGXIII for <u>Telecommunication, Information Industries and Innovation</u> now includes the Information and Telecommunications Technologies Task Force (including ESPRIT and RACE). The Directorate has responsibility for new technologies, information management as well as developments in information technology and telecommunications which have a key role to play in the development and growth of services in the EC. More specifically, DGXIII has instituted plans to stimulate the development of the <u>information market</u> in the Community via improvements in the conditions for growth of the information services sector. Thus a report entitled 'Work Programme for Creating a Common Information Market' (Commission of the European Communities 1985b) was approved by the European Council in 1985. Included under the objectives of this proposal are :

(1) to create internal market conditions which will enable competitive supply of advanced information services in the Community;

(2) to strengthen the Community's position in the growing world information market, from the point of view of suppliers and users alike; and

(3) to ensure that the potential economic, social and <u>regional impacts</u> are fully taken into account in information market initiatives.

A key element in the creation of the Community Information Market (Commission of the European Communities 1985b, paragraph 1a) is that conditions for avoiding a further increase in the gap between information rich and information poor regions of the Community should be implemented. In order to achieve this information poor regions, particularly in respect of SMEs should be in a position to fully participate, through appropriate measures in the development of the Community Information Market. A Senior Official Advisory Group (SOAG) was set up in 1986 to advise DGXIII on developing a Community policy to create a dynamic, common information market in Europe. Finally a major Green Paper on the 'Development of the Common Market for Tele-communications Services and Equipment' has just been produced by DG XIII (Commission of the European Communities 1987).

(9) <u>Regional Policy for Services</u>: Selected service sectors are eligible for funding under the European Regional Development Fund (ERDF), however, as with other Community policy areas, until recently services have not played a prominent part in regional policy in the European Community under the direction of DG XVI for <u>Regional Policy</u>. This is now changing with the realisation of the importance of service activities in the economic development and growth of less favoured regions in Europe (Marquand 1979, 30). In particular the focus is on stimulating those services that are basic (exportable) service activities which can be traded on an inter-regional and international basis, thereby forming part of the export base of a region. As such these types of services have the potential to maintain self-sustaining growth independent of a particular locality and provide significant long term multiplier effects on a local and regional basis. This policy orientation can be seen under the European Regional Development Fund regulations for funding (Council Regulation 1987/84, Article 19) which note that:

"Activities in the service sector qualifying for assistance shall be those concerning tourism or those having a choice of location. Such activities must have an impact on the development of the region and on the level of employment. Tourism activities must contribute to the development of tourism in the region or area in question".

In addition, DGXVI is developing a Community Programme, STRIDE (Science and Technology for Regional Innovation and Development in Europe) which will aim to foster Research and Technological Development (RTD) in the less favoured regions of the Community. A key element in this programme is the role of advanced producer and information services, such as R & D, computer services and technical consultancy, in the development of RTD in such regions.

DG XVI is also involved with a five year programme with DG XIII entitled the Special Telecommunications Action for Regional Development (STAR) programme on the use of advanced telecommunication services to assist in the economic development of the less favoured regions of the Community. Indeed one of the five policy objectives of the European Commission in relation to its telecommunications policy is its commitment "to improve access for the less favoured regions of the Community to benefit from the development of advanced services and networks" (Action Line 4) and this is where the impetus for STAR derives from (Lalor 1987, 115). The STAR programme involves the uprating of the telecommunications system in the following less favoured areas of the Community. Ireland, Greece, Mezzogiorno of Italy, Corsica and the overseas departments of France, Northern Ireland and regions (to be determined) in Spain and Portugal (Regulation COM 85/836 final). The programme more specifically involves the development of advanced telecommunication services in these regions associated with digitalisation to promote rapid introduction of integrated-services digital networks (ISDN), the laying of superimposed networks, especially in high speed data transmission, and the establishment of cellular radio infrastructures. However, of particular importance is that the programme not only focusses on supply related interests but is also concerned with supporting demand stimulation for telecommunication services. They cover promotion exercises, advisory measures, demonstration projects, establishment of service centres, and schemes to promote the take-up of services (especially for SMEs) and the development of specialised regional information services.

<u>Service Policy with Member States</u>: Not only must a services policy for Europe be viewed within the policy environment of the European Commission but also in terms of national government member state policy towards services and local and regional service policy strategies and initiatives. This obviously covers regional policy with a services dimension which has been considerably extended and revised by national governments recently (Tables 7.2 and 7.3). Thus following the 1984 changes in British regional policy services which were of regional importance and would not displace existing jobs (Wood 1984, 285); including such activities as VANS, industrial R&D service, software and data processing became eligible for funding and this has occurred in a number of other EC countries most notably Italy and FR Germany (Table 7.3) but also Luxembourg and the Netherlands. In Italy, under the new Mezzagiorno legislation, and in Germany the emphasis as in the UK, is on helping the development of producer and business services. In Germany more particularly the emphasis has been on regionally exporting service activities ie. those exhibiting what has been termed the 'primary' effect (where more than 50% of sales go outside the region in question). In Italy the purpose of widening the incentive package in the Mezzagiorno was twofold: (1) to encourage firms to <u>use</u> their services and thus improve efficiency and productivity, and (2) to increase the provision and operation of producer services in less favoured areas (Yuill and Allen 1987, 17).

It should also be acknowledged that there are sectoral schemes, aimed at services, which may also involve a regional dimension. In Ireland for example corporation tax for computer service companies has been changed to 10% (Yuill and Allen 1986, 21; see also Norton 1984). The international service division of the Industrial Development Authority (IDA) of Ireland is moreover seeking to make Ireland an international centre for offshore software maintenance services through the attraction of cheap graduate labour and low communication costs which could cut maintenance costs by at least 40% for US companies (Modden 1986). As part of this strategy is the plan to establish an International Financial Services Centre in a redevelopment of Dublin's

TABLE 7.2

MAIN REGIONAL INCENTIVES : ACTIVITY COVERAGE

COUNTRY	INCENTIVE	SERVICES ELIGIBLE FOR REGIONAL AID
Belgium	IS.CG	Producer services, R&D business involved in trade, tourism, management and technical consultancy.
Denmark	IS.MSL	Hotels (but not for MSL) and parts of the transport sector. Hotels are supported only in areas where tourism is very promising.
France	PAT	Research companies and certain tertiary sector activities are eligible in certain designated zones including administration, management, engineering, design and survey activities).
	PRE	Individual regional authorities are free to specify eligible activities. In practice, similar to those under the PAT scheme.
	LBTC	Services are eligible which are not dependent on local markets.
Germany	IA	Regionally exporting services which fulfil the conditions of the primary effect; tourist activities.
	IG	Basically as IA. In addition a special investment grant is available in respect of high grade jobs.
	SDA	All[1]
	ERP	Non-regionally exporting activities with no primary effect.
Ireland	IDA(N).IDA(S)	Services do not normally qualify, but are aided instead through IDA(ISP).
	IDA(ISP)	Target services are: data processing: technical and consultancy services: headquarters: R&D centres: publishing: training services; software development: commercial labs; healthcare services; and international financial services.
Italy	CG.NFSL	Managerial transfers to the South; certain consultancy, research and repair maintenance activities. New Mezzogiorno legislation has extended eligibility to certain producer services.
	SSC	Tourism, commerce, R&D, certain data processing, repair activities, sea transport plus certain producer services.
	TC	Those of an 'industrial' character - certain data processing. R&D repair and handicraft activities.
Luxembourg	CG IS.TC	Service activities may be aided by a CG/IS but only when they make an active contribution to the development of the economy. Certain new technology-related service sector projects are eligible for the TC.
	EL	Hotels and catering are eligible. Other services are eligible only if artisan or commercial - and must involve investment of FLx 500,000.

TABLE 7.2 (continued.../

MAIN REGIONAL INCENTIVES : ACTIVITY COVERAGE

Netherlands	IPR	Regionally exporting services plus certain research departments if of importance to industrial activities. Large 'footloose' tourist projects qualify, but not hotels/restaurants.
United Kingdom (a) Great Britain	RDG	Certain repair and maintenance activities; scientific R&D and training related to qualifying activities are eligible. Also eligible are industrial R&D services; banking, finance and insurance services; business services; mail order houses; football pools and various transport and communication services.
	RSA	No SIC Orders are explicitly excluded[2]. However, purely local, consumer-type services would not be eligible. Awards are very much concentrated on the manufacturing sector.
(b) Northern Ireland	SCG	Services are not generally eligible since the SCG is restricted to sectors traded or tradeable beyond Northern Ireland.
	SA	Regionally-exporting services, services which fill a genuine niche in the local market, services where the overall impact is to increase employment.
Greece	IG.IRS.IDA.TA	Tourism and within the transport sector, certain coastal shipping lines are eligible. Certain repair activities are explicitly excluded.
Portugal	FA.TC	None.
Spain	RIG.POC	Only service sector projects put forward by private firms in the fields of education, health and tourism.

see separate Abbreviation List.

(1) Not activity-related.
(2) Although no SIC Orders are explicitly excluded under the scheme, awards are very much concentrated on the manufacturing sector.

Source: Compiled from Allen and Yuill (1987, 87-8).

TABLE 7.3

MAIN REGIONAL INCENTIVES : CHANGES IN ACTIVITY COVERAGE

COUNTRY	CHANGE
BELGIUM	No change.
DENMARK	A number of minor amendments have been made to the list of activities excluded because they were operating under 'strained' market conditions.
FRANCE	No explicit change change. However, budgetary constraints have led to a greater selectivity of award. As a result, projects from computing or data-processing firms are now virtually systematically rejected for PAT awards.
FR GERMANY	The list of eligible activities for the IA and IG has been extended to include a range of new business services (eg. market research, management and technical consultancy, laboratory services, industrial R&D, advertising and exhibition facilities). In addition, the IG may be awarded to firms in sectors not on the eligibility list where it can be demonstrated (on an individual case basis) that the goods or services produced are regionally exported and there is a significant net local benefit. These measures reflect an increased emphasis on the service (and crafts) sector. Also in Germany, support for innovative activities and 'human' capital has increased – through higher IG rates for the creation of high-grade jobs; through including 'intangible' assets within eligible expenditure; through easing eligibility conditions for expanding enterprise centre projects; and through allowing cumulation of regional and R&D investment allowances.
IRELAND	No change. However, the 1986 Industrial Development Act gives scope for introducing new mechanisms to promote development, including the offer of assistance towards the cost of obtaining licences or paying royalty charges.
ITALY	Eligibility for the capital grant, soft loan and social security concession has been extended for the first time beyond manufacturing to include certain producer services (designated by the CIPI). In similar vein, eligible expenditure for the capital grant/soft loan has been widened beyond fixed investment to include the purchase of licences, the setting up of offices and the creation of distribution networks for Mezzogiorno products. Finally, the CIPI has revised the list of (manufacturing) activities for which eligibility has been suspended and also the list of priority sectors eligible for awards one-fifth above the basic rate.
LUXEMBOURG	The new Industrial Framework Law has extended tax concession eligibility to include new technology-related service sector projects.
NETHERLANDS	Following a 1984–86 experiment regarding their eligibility, 'footloose' tourist projects are now eligible for the IPR. Hotels and restaurants are, however, explicitly excluded.
UNITED KINGDOM	In Northern Ireland, eligible sectors under the SCG scheme must now generally be traded or tradeable beyond Northern Ireland. As a result, quarrying, the manufacture of concrete products, bakeries and the printing and publishing of newspapers are no longer eligible. Similarly, services are generally not eligible for SCG assistance.
GREECE	Tourist ships built in Greek yards are now eligible for assistance. On the other hand, a number of repair activities are now explicitly excluded.
PORTUGAL	No change.
SPAIN	No change.

see separate Abbreviation List

Source: Yuill and Allen (1987, 85).

ABBREVIATION LIST FOR TABLES 7.2 AND 7.3

BELGIUM : IS - interest subsidy; CG - capital grant.

DENMARK : CSL - company soft loan; IG - investment grant; MSL - municipality soft loan; GDR - General Development Regions; SDR - Special Development Regions.

FRANCE : PAT - regional policy grant; PRE - regional employment grant; PDR - regional development grant; PLAT - service location grant; LBTC - local business tax concession; FSAI - special fund for industrial adaptation.

GERMANY : IA - investment allowance; IG - investment grant; SDA - special depreciation allowance; ERP - European Recovery Programme regional soft loan; GA - Gemeinschaftsaufgabe; ZBA - Zonal Border Area.

IRELAND : IDA - Industrial Development Authority; IDA (N) - IDA grant (new industry); IDA (RM) - IDA grant (re-equipment and modernisation); IDA (S) - IDA grant (small industries programme).; IDA(ISP) - IDA international services programme; DA - Designated Areas; NDA - Non-Designated Areas.

ITALY : CG - capital grant; NFSL - national fund soft loan; SSC - social security concession; TC - tax concession.

LUXEMBOURG : CG - capital grant; IS - interest subsidy; TC - tax concession; EL - equipment loan; SCNI - National Investment Corporation.

NETHERLANDS : IPR - investment premium; WIR(RA) - investment account (regional allowance).

UNITED KINGDOM : RDG - regional development grant; RSA - regional selective assistance; RSA(G) - RSA grant; RSA(L) - RSA soft loan; OSIS - office and service industries scheme; SDA - Special Development Areas; DA - Development Areas; IA - Intermediate Areas; SCG - Northern Ireland standard capital grant; SA - Northern Ireland selective assistance; CTRG - corporation tax relief grant.

GREECE : IG - investment grant; IRS - interest rate subsidy; IDA - increased depreciation allowance; TA - tax allowance.

PORTUGAL : FA - financial assistance; TC - tax concessions.

SPAIN : RIG - regional investment grant; POC - priority in obtaining official credit.

170

Custom House Docks. The scheme would aim to attract the range of international finance activity : offshore financing, foreign exchange trading, futures, Eurobond and other securities dealing, leasing, insurance and reinsurance inhouse treasury management and back-office operations such as data processing. The strategy would seek to attract banks and securities houses away from the other major European financial centres, in particular London. Already US and Japanese banks together with members of the Chicago Mercantile Exchange are supposed to be involved in the scheme. Equally a wide and diverse number of local and regional initiatives are being implemented by local authorities and regional governments or agencies in member states (see, for example, Dineen 1986). Any long term policy strategies for services in the Community therefore has to be consistent with these schemes.

Services and Economic Development in the Regions : Long Term Policy Issues

The final section here raises some key issues which need to be considered for a long term strategy to develop service, in particular, information service, activity in the regions of Europe. Information and producer services are becoming increasingly integrated on a global scale, which is in turn leading to the gradual breakdown of the sub-national and regional orientation of service activity. As such regions are becoming more 'open' in an information service market context. Associated with this trend are linked other processes, such as, the blurring of service sectors both between themselves and with manufacturing. This is in turn related to diversification by firms and the increasing interweaving role that information services are playing in industrial economies.

Other issues are also relevant here. Thus individual nations are seeking a competitive advantage on the globalisation of service activities via policy initiatives and changes in the regulatory framework (Chapter 3). However there is relatively little or no effective control by national or supra-national bodies on these globalisation shifts in services. The exception have been the international technocratic bodies, such as the International Telegraph and Telephone Consultative Committee (CCITT) of the International Telecommunications Union (ITU) which are making international technical agreements that are having far-reaching (often unjudged) economic impacts on the growth and development of information and other services. These international technocratic bodies have helped set the parameters of international service growth and development.

Associated with the development towards a global market has been the role of multinational corporations which have been largely responsible for the growth of transborder data flows on both an inter- and intra-organisational level. They in turn have created their own international pressure groups, such as the International Telecommunications Users Group (INTUG) and the International Data Exchange Association (IDEA), to get their own policy requirements known and implemented. Clearly all these multi-faceted trends and processes in the globalisation of information and other services emphasises the importance of long term strategic thinking in this area. As such consideration should be given to providing more help in setting up non-European collaborative ventures in advanced information and producer services. Indeed a number of European information service companies are already developing joint partnerships and ventures: for example, ESA-IRS based in Frascati and Pergamon Infoline based in London have agreed to an interconnection which allows customers of each company to access the others' database. Similarly Excerpta Medica owned by Elsevier and Data-Star of the UK are developing a joint medical database; whilst UK Tandata Holdings and ITT Netherlands are developing an interactive television cable network in the Netherlands.

As noted earlier, regions are also faced with increasing concentration levels in services and manufacturing. In terms of less favoured regions this increasing sectoral concentration is accompanied by increasing spatial concentration in control and decision making in the core regions of the Community. This is related to the 'branch plant economy' syndrome where service establishments in the more peripheral areas of the European Community will become associated with less sophisticated, routine services controlled from headquarters units located in core metropolitan regions. However again within this process are other linked themes: the gradual breakdown of individual service sectors associated with diversification; the commodification and externalisation of services; changes in the regulatory environment; and the internationalisation of information and other service markets. The competitive conditions of many service markets are therefore being increasingly dominated by giant corporations, which will have implications for entry barriers and competition. The trend towards concentration in information and other service markets therefore has implications for SME policy in these sectors. More particularly, however, it has implications for regions in terms of autonomy and dependence over the development of services in their locality.

Privatisation, deregulation and changes in the nature and level of public ownership in services similarly, create both considerable threats and opportunities to the less favoured regions of Europe. A major issue, which requires further research, revolves around universality and cross subsidisation in telecommunications (and transport). Other issues cover the consequences of 'opening up' service markets both on a national and regional basis, in terms of the survival and growth of firms. Thus the opening up of service markets may be best for consumers but may have profound negative consequences for smaller, less efficient and innovative service producers, especially found in the peripheral, less favoured regions of Europe. Again this may entail substantial job losses in the less favoured regions and strengthen the tendency for these localities to become 'branch plant economies' in services with employment being in primarily low skilled, part-time jobs. There is also the impact of service deregulation and liberalisation in the supporting manufacturing base of the regions. New externally controlled service establishments are likely to maintain input linkages for goods with suppliers located in the home market. Lastly there is the issue of the importance, indeed dominance, of public sector service employment in many less favoured regions of Europe (which involves a substantial net transfer payment to these regions). A reduction in national expenditure on public services therefore may have a major differential economic and employment impact on these localities.

Finally, technological innovation and the development of telecommunication networks is having a profound impact on the growth and development of services, in particular information services. However, key issues such as access to telecommunication networks, entry barriers in the fields of technological innovation in respect of information services and size factors all have relevance to the participation of less favoured regions in information services and new service markets.

Of concern here is the ability of industry in less favoured regions to have access to public telecommunication networks, in particular advanced telecommunication facilities, and entry costs of privately provided communication network systems. Moreover, many private intra-corporate communication networks although located in less favoured regions are not accessible by outside firms. This is not to suggest that it is a simple supply led, infrastructure problem. Some of the key problems in less favoured regions in relation to networks and advanced information services relate to demand aspects; the type and nature of firms in the region; corporate behaviour and organisational structure; and the leakage of information demand to core regions.

7 The information-based service economy in Europe: conclusions

Information services have become a key element in the growth and development of the European economy. It has been outlined that information services are of economic importance for a number of strategic reasons. Information services represent key economic and employment growth sectors in their own right (Chapter 4). In addition they form major growth components within other industries. Associated with this on a macro-economic level, information services represent a key integrating and facilitating activity across the spread of industrial activity. On a micro-economic, intra-organisational basis the efficient generation and distribution of information and data between geographically dispensed sites has provided a key role in the development of large multinational corporations. On an inter-organisational level the evolutional growth of private corporate computer-communication networks is also now leading to the more effective flow of information between corporate partners in terms of joint ventures and other collaborative projects.

Information networks also represent a pivotal element in the competitive structure of modern industrial economies in terms of their control function (Chapter 5). A recent illustration of this is the use of computerised reservation and distribution systems (CRS) to widen an airline's sphere of influence by drawing in smaller airlines, travel agents, hotels and car hire companies to use and become dependent on their information network. The key competitive edge in the airline industry has become focussed on developing a large and sophisticated CRS information system. Thus in Europe US airlines such as American Airlines with its Sabre system and United Airlines with its Apollo system, are pushing their sophisticated reservation systems to encroach on the European airline market.

Lastly information services are increasingly seen as a key element in foreign trade and investment. Again this is not only in terms of their direct trade value, which is still relatively small compared with manufactured goods, but also in relation to their facilitating role and in a 'combinational' sense (for example, the quality of computer software plays a major role in computer hardware sales). Up until recently however the real significance of information services to researchers and policymakers has not been recognised. A rapid change in attitude needs to take place in Europe if it is not to loose out in terms of growth, competitive advantage and control of its economy.

Key Processes in Information Services Development

Chapter 3 outlined the major processes that were influencing the development of information services in the Community. Developments in technology are obviously having a profound effect on those information services that are emerging and in how existing services are distributed and used (Chapter 6). The internationalisation of information services is associated with this, as communication networks have grown allowing wider spatial market opportunities. A key force in this development has been the widening role of giant multinational service corporations and conglomerates operating across a range of geographical and sectoral markets. Deregulation of service markets by national governments has been an important factor in facilitating this together wider trade liberalisation associated with the current round of GATT negotiations and, on a Community level, the programme being undertaken by the Commission of completing the Internal Market.

The rapid growth and technological development of information service markets together with the evolving economic structure of such newly emerging industries (Chapter 5) has led to increasing concentration in many information sectors. However in conjunction with this niche market opportunities have been created by these (often rapidly) expanding and evolving market systems for innovative and dynamic SMEs. New sectors in many of these newly emerging information industries have not come via the appearance of new firms but rather have originated from existing companies diversifying into these sectors through the process of externalisation or acquisition (Chapter 3).

The Location of Information Services and Economic Growth

A final but important interweaving element in the development of information services in Europe is that of location. A number of key processes outlined earlier that are associated with the development of information services are highly influenced by geography. The growth of public telecommunications and computer-communication networks are above all geographically delimitated. The configuration, level of access and quality of these networked information services are all set within a clear geographical framework. Moreover the technological development and upgrading of these network systems used by information services have clear-cut, core-periphery and urban-rural patterns. In addition the introduction and flows of new information technology and service developments also have distinct urban and regional contrasts favouring the core metropolitan areas of national territories. Above all information generation and diffusion has a clear spatial dimension which favours large urban centres and core regions.

Equally international trade in information and services is set within an important locational framework. Thus the trade performance of the UK economy in terms of information and business related services is profoundly influenced by the continued development of London as international financial and trading centre, set with its network of global information and communication links and its supporting infrastructure. The growth and development of service trade, associated with trends in internationalisation and liberalisation, is going to have a substantial geographical impact in terms of economic growth between cities and regions within national territories and wider trading blocs such as the European Community. It was also noted that deregulation policies were being used to enhance the locational attractiveness in terms of foreign investment and trade in information services and infrastructure of 'national champion' cities.

A further spatial dimension, associated with trade liberalisation and the opening of national economies, is the need to geographically extend the information service market so that it is of sufficient size for the full realisation of technological development and competitive efficiencies to be gained via scale economics associated with large homogenous markets. A key element therefore in the Completion of the Internal Market in information services in the European is that it will break down geographical sub-markets within European and will create instead a single, open market for European service companies.

Finally, just as there are market spatial variations in the opportunities for information service growth in terms of markets and employment so there will be areas or regions where these opportunities will be lacking or highly constrained. As such policies need to be designed to overcome the interwoven cumulative technological, organisational and structural constraints that are restricting the spread and indigenous development of information activities in these localities.

References

AARONOVITCH, S. and SAMSON, P. (1985) The Insurance Industry in the
  Countries of the EEC : Structure, Conduct and Performance.
  Commission of the European Communities, Luxembourg.

ABLER, R. and ADAMS, J.S. (1977) 'The industrial and occupational
  structure of the American labor force'. Papers in Geography 15,
  Department of Geography, Pennsylvania, State University.

ADR/EMPIRICA/TAVISTOCK INSTITUTE (1985) Distance Working in the Federal
  Republic of Germany, France and the United Kingdom, FAST Distance
  Working Paper No.2, Tavistock Institute, London.

ANTONELLI, C. (1984) "Multinational firms, international trade and
  international telecommunications" Information Economics and Policy
  2, 333-43.

ANTONELLI, C. (1985) "The diffusion of an organisational innovation:
  international data telecommunications and multinational industrial
  firms" International Journal of Industrial Organisation 3, 109-11.

ARMSTRONG, R.B. (1972) The Office Industry : Patterns of Growth and
  Location. MIT Press, Cambridge, Mass.

ARMSTRONG, R.B. (1979) 'National trends in office construction,
  employment and headquarters location in US metropolitan areas' in
  Daniels, P.W. (ed.) Spatial Patterns of Office Growth and Location,
  Bell, London, 61-94.

ARNOLD, E. (1983) 'Office Automation and Employment' in Turney, J.
(ed.) Science and Technology Worldview, Pluto Press, London.

ARNOLD, E., BURKE, L. and FAULKNER, W. (1981) "Women and micro-
electronics : the case of word processing", Womens Studies
International Quarterly Vol.4, No.3.

ARONSON, J.D. and COWHEY, P.F. (1984) Trade in Services : A Case for
Open Markets American Enterprise Institute for Public Policy
Research, Washington.

ARTHUR ANDERSON and CO. (1986) The Decade of Change : Banking in Europe
- The Next 10 Years, Lafferty Publications, London.

ASH, N.I. (1986) "On-line databases", Money Management December 1986,
45-48.

ASSOCIATION OF DATA PRODUCERS (1986) The British Database Industry,
ADP, London.

ATKINSON, J. (1985) "Flexibility: planning for an uncertain future"
Manpower Policy and Practice 1, 26-29

AYDALOT, P. (1986) 'The location of new firm creation : the French
case' in Keeble, D.E. and Wever, E. (Eds.) New Firms and Regional
Development in Europe, Croom Helm, London, 105-123.

BADE, F.J. (1986) 'The economic importance of small and medium-sized
firms in the Federal Republic of Germany', in Keeble, D.E. and
Wever, E. (Eds.) New Firms and Regional Development in Europe,
Croom Helm, London, 256-274.

BAILLY, A.S. (1985) 'Le Secteur des Services : Une Chance pour le
Developpement Local' in The Present and Future Role of Services in
Regional Development : 16-18 October 1985, Seminar Proceedings -
External Contributions, Occasional Paper No.74, FAST, Brussels.

BAKIS, H. (1980) The communications of larger firms and their
implications on the emergence of a new industrial order.
Contributing Report : Commission of Industrial Systems -
International Geographical Union, Chuo University, Tokyo, 26-30
August.

BAKIS, H. (1982) The geographical impact of telecommunication systems
used within firms. Paper presented at meeting of Commission on
Industrial Systems-International Geographical Union. Universidade
de Sao Paulo, 5-15 August.

BAKIS, H. (1985) Telecommunication and organisation of company work
space. Paper presented at meeting of Commission on Industrial
Systems - International Geographical Union. Universiteit Nimegen,
19-24 August.

BANNON, M.J. (1985) 'Service activities in national and regional
development : trends and prospects for Ireland' in Services - The
New Economy : Implications for National and Regional Development,
Proceedings of the Irish Branch of the Regional Studies Association
1985 Annual Conference, 26 March, 1985, Dublin.

BANNON, M.J. and BLAIR, S. (1985) Service Activities, The Information Economy and the Role of Regional Centres, EIE Publications, Dublin.

BARAN, B. (1985a) Technological Innovation and Deregulation : The Transformation of the Labor Process in the Insurance Industry. BRIE Working Paper, University of California, Berkeley.

BARAN, B. (1985b) "Office automation and women's work : the technological transformation of the insurance industry in Castells, M. (Ed.) High Technology, Space and Society, Sage Publications, Beverly Hills 143-171.

BARRAS, R. (1984a) Information Technology and Economic Perspectives : The Case of Office Based Services. Technical Change Centre Report, London.

BARRAS, R. (1984b) Growth and Technical Change in the UK Service Sector Technical Change Centre Report, London.

BARRAS, R. (1985a) Technical Tools for the New Services, Draft Final Report for FAST II, Directorate General for Science, Research and Development, Commission of the European Communities, Brussels.

BARRAS, R. (1985b) "Information technology and the service revolution", Policy Studies 5, 14.

BARRAS, R. (1986) "Towards a theory of innovation in services" Research Policy 15, 161-173.

BARRAS, R. and SWANN, J. (1983) The Adoption and Impact of Information Technology in the UK Insurance Industry. Technical Change Centre Report.

BARRAS, R. and SWANN, J. (1984a) The Adoption and Impact of Information Technology in the UK Accountancy Profession. Technical Change Centre Report.

BARRAS, R. & SWANN, J. (1984b) 'Information technology and the service sector : quality of services and quality of jobs' in Marstrand, P. (Ed.) New Technology and the Future of Work and Skills, Frances Pinter.

BARTELS, C.P.A., BOONSTRA, D. and VLESSERT, H.H. (1983) Prospects for the Regional Development of the Tertiary Sector, Buro Bartels, Oudmelon, The Netherlands for Directorate General for Regional Policy, Commission of the European Communities.

BARTELS, C.P.A., WERKHOVEN, F.G.M. and KRUIJK, P.D. (1984) Employment in Retail Trade in EC Countries. Buro Bartels, Oudmelon. The Netherlands for Directorate General of Employment, Social Affairs and Education, Commission of the European Communities.

BEARSE, P.J. (1978) "On the intra-regional diffusion of business service activity", Regional Studies 12, 563-578.

BELL, D. (1974) The Coming of Post-Industrial Society, Heinemann, London.

BERRY, B.J.L. (1972) 'Hierarchical Diffusion' in Hansen, N.M. (Ed.) 'Growth Centers in Regional Economic Growth', Free Press, New York, 108-138.

BEYERS, W.B. and ALVINE, M.J. (1985) "Export services in post-industrial society". Papers of the Regional Science Association 57, 33-45.

BEYERS, W.B., ALVINE, M.J. and JOHNSON, E.G. (1985) The Service Economy :Export of Services in the Central Puget Sound Region. Central Puget Sound Economic Development District, Seattle, Washington.

BEYERS, W.B., TOFFLEMIRE, J.M., STRANAHAN, H.A. and JOHNSON, E.G. (1986) The Service Economy : Understanding Growth of Producer Services in the Central Puget Sound Region, Central Puget Sound Economic Development District, Seattle, Washington.

BEVAN, S. (1984) Secretaries and Typists : The Impact of Office Automation, Institute of Manpower Studies, Report No.93, Brighton.

BLEAZARD, B. (1986) Value Added Network Services: State of the Art Report National Computing Centre Ltd, Manchester.

BORCHERT, J.R. (1978) "Major control points in American economic geography" Annals of the Association of American Geographers 68, 214-32.

BRODIE, I. (1986) 'Distributive Trades' in Smith, A.D. (ed.) Technological Trends and Employment : 5. Commercial Service Industries, Gower, Aldershot, 121-198.

BROWN, L.A. & MALECKI, E.J. (1977) "Comments on landscape evolution". Regional Studies 11, 211-223.

BRUCE, E.R., CUNARD, J.P. and DIRECTOR, M.D. (1986) From Telecommunications to Electronic Services, Butterworths, London.

BURNS, L.S. (1977) "The location of the headquarters of industrial companies", Urban Studies 14, 211-15.

CANADIAN ECONOMIC SERVICES LIMITED (1977) Issues in the Analysis of the Information Sector of the Canadian Economy, Ottawa, 214.

CAPPELLIN, R. (1985) 'The Development of Service Activities in the Italian Urban System in The Present and Future Role of Services in Regional Development : 16-18 October 1985, Seminar Proceedings - External Contributions, Occasional Paper No.74, FAST, Brussels.

182

CAPPELLIN, R. and GRILLENZONI, C. (1983) "Diffusion and specialisation in the location of service activities in Italy" Sistemi Urbani 2, 249-282.

CENSIS (1985) Business Services As An Industrialisation Variable in Underdevelopment Regions : 2 Case Studies in Southern Italy, Centro Studi Investimenti : Sociali, Roma. Report for FAST II, Directorate General for Science, Research and Development, Commission of the European Communities, Brussels.

CHANNON, D.F. (1978) The Service Industries : Strategy, Structure and Financial Performance, Macmillan, London.

CHILD, J., LOVERIDGE, R., HARVEY, J. and SPENCER, A. (1984) 'Microelectronics and the quality of employment in services', in Marstrand, P. (Ed.) New Technology and the Future of Work and Skills, Frances Pinter, London.

CLAIRMONTE, F. and CAVENAGH, J. (1984) "Transnational Corporations and services : the final frontier" Trade and Development 5, 215-273.

COMMISSION OF THE EUROPEAN COMMUNITIES (1984a) Fast, 1984-1987 Objectives and Work Programmes, Commission of the European Communities, Brussels XII-201-84-EN.

COMMISSION OF THE EUROPEAN COMMUNITIES (1984b) The Regions of Europe : Second Periodic Report on the Social and Economic Situation of the Regions of the Community. Commission of the European Communities, Luxembourg.

COMMISSION OF THE EUROPEAN COMMUNITIES (1985a) Completing the Internal Market. White Paper from the Commission to the European Council, COM(85) 310 final, 28-9, June, Milan.

COMMISSION OF THE EUROPEAN COMMUNITIES (1985b) Work Programme for Creating a Common Information Market, Communication from the Commission to the Council, COM(85) 658 final, 29 November, Brussels.

COMMISSION OF THE EUROPEAN COMMUNITIES (1987) Towards A Dynamic European Economy : Green Paper on the Development of the Common Market for Telecommunication Services and Equipment, Communication by the Commission of the European Communities, Brussels, COM(87) 290 final.

CORDELL, A.J. (1985) The Uneasy Eighties : The Transition to an Information Society. Science Council of Canada, Background Study 53.

CROMPTON, R. and JONES, G. (1984) White-Collar Proletariat : Deskilling and Gender in Clerical Work, Macmillan, London.

CRUM, R.E. and GUDGIN, G. (1977) Non-production activities in UK manufacturing industry. Regional Policy Series 3, Collection Studies, Commission of the European Communities, Brussels.

CSP INTERNATIONAL LTD (1986) The Global Structure of the Electronic Information Services Industry British Library Research Paper 1, British Library, Boston Spa.

CUADRADO, J.R. (1986) Supply and Demand of Services and Regional Development : The Case of Comunidad Valenciana (Spain), FAST Occasional Paper No.93.

CULLEN, A. (1986) "Electronic information services : an emerging market opportunity?" Telecommunications Policy 10, 299–312.

d'ALCANTARA, G. (1986) Concepts for the Improvement Measurement and Formalisation of Productivity in the Services : Parts 1 and 2, Final Report for FAST II, Directorate General for Science Research and Development, Commission of the European Communities, Brussels, FAST Occasional Paper No.95 A and B.

DAMESICK, P.K. (1986) "Service industries, employment and regional development in Britain : a review of recent trends and issues". Transactions Institute of British Geographers, New Series 11, 212–226.

DANIELS, P.W. (1984) "Modern technology in provincial offices : some empirical evidence". Service Industries Journal 3, 21–41.

DANIELS, P.W. (1985) Service Industries : A Geographical Appraisal, Methuen, London.

DANIELS, P.W. (1986a) 'Producer services in the UK space economy' in Martin, R. and Rowthorne, R. (Eds.) The Geography of Deindustrialisation, Macmillan, London, 291–321.

DANIELS, P.W. (1986b) "Foreign banks and metropolitan development: a comparison of London and New York" Tijdschrift voor Economische en Sociale Geographie 77, 269–287.

DESIATA, A. (1986) 'Financial innovation, and the development of a European financial market' in Commission of the European Communities, Symposium on Europe and the Future of Financial Services: Proceedings/Communications, DGXII (FAST) – DGXV, Brussels, 83–97.

DEWHURST, J.H. (1985) Output, employment and labour productivity in the Scottish service sector, 1963–1980. Paper presented to British Annual Conference of the Regional Science Association, Manchester, 4–6 September.

DINEEN, D.A. (1986) Strategy for Development of the Services Sector in the Mid-West Region of Ireland Mid-West Regional Development Organisation, Limerick.

DINTEVEN, J.H.J. van (1987, forthcoming) "The role of business service offices in the economy of medium-sized cities" Environment and Planning A 19.

DOKOPOULOU, E. (1986) 'Small manufacturing firms and regional development in Greece : patterns and changes' in Keeble, D.E. and Wever, E. (Eds.) New Firms and Regional Development in Europe, Croom Helm, London, 299-317.

DORDICK, H.S., BRADLEY, H.G., NANUS, B. and MARTIN, T.H. (1979) "Network information services - the emergence of an industry". Telecommunications Policy 3, 217-234.

DORDICK, H.S., NANUS, B. and BRADLEY, H.G. (1981) The Emerging Network Marketplace Ablex, Norwood, N.J.

DOSWELL, A. (1983) Office Automation, Chichester : John Wiley.

DUCKER, J. (1985) "Electronic information-impact of the database" Futures 17, 164-169.

DUNN, D.A. (1978) Limitations on the growth of computer - communication services. Telecommunications Policy 2, 106-116.

DUNNING, N.J. and NORMAN, P. (1979) 'Factors Influencing the Location of Offices of Multinational Enterprises. Location of Offices Bureau Research Paper No.8, Economists Advisory Group, London.

DUNNING, J. and NORMAN, P. (1983) "The theory of the multinational enterprise : an application to office location" Environment and Planning A 15, 675-692.

DYSON, K. (1986) 'West European States and the Communications Revolution' in Dyson, K. and Humphreys, P. (Eds) The Politics of the Communications Revolution in Western Europe Frank Cass, London, 10-55.

EIU/CURDS (1985) Availability, Cost and Use of Telecommunications in the Northern Region. Report to NECCA, NEDC, DTI by Economist Informatics Unit and Centre for Urban and Regional Development Studies, University of Newcastle upon Tyne.

ESTRIN, D.L. (1985) "Inter-organisational networks : stringing wires across administrative boundaries", Computer Networks and ISDN Systems 9, 281-295.

ESTRIN, D.L. (1986) Access to Inter-Organisation Computer Networks (mimeo) Department of Electrical Engineering and Computer Science, Massachusetts Institute of Technology, Cambridge, Mass.

EVANS, A.W. (1973) "The location of the headquarters of industrial companies" Urban Studies 10, 387-395.

FASE, M.M.G. (1984) Informele economie en geldomloop in De Informele Economie. Preadviezen van de Vereniging voor de Staathuishoud Kunde, Leiden.

FEKETEKUTY, G. (1986) "Impact on informatics and communications and their free flow" Telecommunication Journal 53, 589-595.

FELGRAN, S.D. and FERGUSON, R.E. (1986) "The evolution of retail EFT networks" New England Economic Review July/August 1986, 42-56.

FERTIG, R.T. (1985) The Software Revolution North Holland, New York.

FIRN, J.R. (1975) "External control and regional development : the case of Scotland" Environment and Planning A 7, 393-414.

FITZPATRICK, J. (1985) 'Technology and economic development : the role of private services' in Services - The New Economy : Implications for National and Regional Development. Proceedings of the Irish Branch of the Regional Studies Association 1985 Annual Conference, 26 March 1985, Dublin.

FITZPATRICK, J. and McERIFF, J. (1984) Technical Change and International Competitiveness in Service Industry : Implications for Ireland, EIE Publications, Dublin.

FLETCHER, J. and SNEE, H. (1985) "The need for output measurements in the service industries : a comment" The Service Industries Journal 5, 73-78.

FLOWERDEW, A.D.J., OLDMAN, C.M. and WHITEHEAD, C.M.E. (1984) The Pricing and Provision of Information : Some Recent Official Reports, Library and Information Research Report 20, British Library Lending Division, Boston Spa.

FOGELMAN, T. (1985) Technical Aspects of Distance Working, FAST Distance Working Paper No.4, Tavistock Institute, London.

FOORD, J. and GILLESPIE, A.E. (1985) Reorganisation, New Technology and Office Jobs. CURDS Discussion Paper No.75, University of Newcastle upon Tyne.

FREY, B.S. and WECK-HANNEMAN, H. (1984) "The hidden economy as an 'unobserved' variable" European Economic Review 26, 33-53.

FROST and SULLIVAN INC. (1985) Value Added Networks in Europe. Frost & Sullivan, London.

FROST and SULLIVAN INC. (1987) The West European Market for Value Added Network Services Frost and Sullivan, London.

GATT (1985) Services : First Analytical Summary of Information Exchanged Among Contracting Parties. General Agreement on Tariffs and Trade (GATT), April.

GERSHUNY, J. (1985) 'Services and innovation : the future for employment' in Services - The New Economy : Implications for National and Regional Development. Proceedings of the Irish Branch of the Regional Studies Association 1985 Annual Conference, 26 March, Dublin.

GERSHUNY, J. and MILES, I. (1983) The New Service Economy, Frances
    Pinter, London.

GILLESPIE, A.E., GODDARD, J.B., ROBINSON, J.F., SMITH, I.J. and
THWAITES, A.T. (1984) 'The effects of new information technology on the
    less-favoured regions of the Community', Studies Collection,
    Regional Policy Series 23, Brussels.

GILLESPIE, A.E. and HEPWORTH, M.E. (1986) 'Telecommunications and
    Regional Development in the Information Society' Newcastle Studies
    of the Information Economy, Working Paper No.1, CURDS, University
    of Newcastle upon Tyne.

GLEAVE, D. (1985) The Impact of Innovations on Service Employment
    (mimeo), The Technical Change Centre, London, 22.

GODDARD, J.B. (1975) Office Location in Urban and Regional Development,
    Oxford University Press, Oxford.

GODDARD, J.B. (1977) "Urban geography : city and regional systems".
    Progress in Human Geography 1, 296-303.

GODDARD, J.B. and SMITH, I.J. (1978) "Changes in corporate control in
    the British urban system, 1972-1977". Environment and Planning A
    10, 1073-1084.

GOTTMAN, J. (1970) "Urban centrality and the inter-wearing of
    quaternary functions" Ekistics 29, 322-331.

GREEN, A.H. (1981) 'Transfer pricing, its relatives and their control
    in developing countries : notes towards an operational definition
    and approach', in Murray, R. (Ed.) Multinationals Beyond the
    Market: Intra-Firm Trade and the Control of Transfer Pricing, The
    Harvester Press, Brighton, 221-244.

GREEN, M. (1985) "The development of market services in the European
    Community, the United States and Japan" European Economy 25, 69-96.

GREEN, M. (1986) 'Updated statistics on the role of service activities
    in the Community Economy'. Paper circulated to the Interservice
    Group, Commission of the European Communities, Brussels.

GUDGIN, G. (1983) Job Generation in the Service Sector. Department of
    Applied Economics, University of Cambridge. Position paper for the
    Industry and Employment Committee of the SSRC, London.

GUERIN, A. and OUTREQUIN, P. (1986) Innovations Sociales daus les
    Services et Developpement Regional en Lorraine. Final Report for
    FAST II, Directorate General for Science, Research and Development,
    Commission of the European Communities, Brussels.

GURNSEY, J. (1986) "Electronic publishing : a state-of-the-art review"
    Information Media & Technology 18, 101-104.

HAKIM, C. (1984) "Homework and outwork" Employment Gazette 92:1, 7-12.

HALL, C. (1983) 'The Federal Republic of Germany', in Storey, D.J. (Ed.) The Small Firm : An International Survey, Croom Helm, London, 153-178.

HAYWOOD-FARMER, J. and NOLLETT, J. (1985) "Productivity in professional services" The Service Industries Journal 5, 169-180.

HEPWORTH, M.E. (1986a) "The geography of technological change in the information economy" Regional Studies 20, 407-424.

HEPWORTH, M.E. (1986b) "The geography of economic opportunity in the information society" The Information Society 4, 205-220.

HEPWORTH, M.E. (1987, forthcoming) "Information services in the international network marketplace" Information Services and Use 7.

HEPWORTH, M.E., GREEN, A.E. and GILLESPIE, A.E. (1987) "The spatial division of information labour in Great Britain" Environment and Planning A 19, 793-806.

HERRMANN, A., OCHEL, W., WEGNER, M., BARRAS, R. and PETERSON, J. (1985) The International Competitiveness of the European Communities' Tradeable Service Sector - A Feasibility Study Report to Directorate General for Internal Market and Industrial Affairs, Commission of the European Communities, Brussels.

HEWLETT, N. (1985) "New technology and banking employment in the EEC" Futures 17, 34-44.

HOLLAND, P. (1985) "Banking in the eighties : adaptation or innovation?" The Service Industries Journal 5, 215-225.

HOLTI, R. and STERN, E. (1984) Social Aspects of New Information Technology in the UK, Tavistock Institute of Human Relations, London, 2-T 485.

HOLTI, R. and STERN, E. (1985) The Origins and Diffusion of Distance Working FAST, Distance Working Project, Working Paper 3, Tavistock Institute of Human Relations, London.

HOCKIN, N.M.C. (1978) "Data networks for business and government". Telecommunications Policy 2, 117-127.

HOWELLS, J. (1984) "The location of research and development : some observations and evidence from Britain". Regional Studies 18, 13-29.

HOWELLS, J. (1987a) Technological Innovation, Industrial Organisation and Location of Services in the European Community: Regional Development Prospects and the Role of Information Services Final Report for FAST II, Directorate General for Science, Research and Development, Commission of the European Communities, Brussels. FAST Occasional Paper No.142.

188

HOWELLS, J. (1987b, forthcoming) "Developments in the location, technology and industrial organisation of computer services : some trends and research issues", Regional Studies 21.6.

HOWELLS, J. (1987c, forthcoming) 'International Trade in Information Services : Implications for National and Regional Development' Newcastle Studies of the Information Economy, Working Paper No.5, CURDS, University of Newcastle upon Tyne.

HOWELLS, J. and GREEN, A. (1986a) Location, Technological Innovation and Structural Change in UK Services : National and Regional Economic Development Prospects. Final Report for FAST II, Directorate General for Science, Research and Development, Commission of the European Communities, Brussels. FAST Occasional Paper No.101.

HOWELLS, J. and GREEN, A. (1986b) "Location, technology and industrial organisation in UK services", Progress in Planning 27, 83-184.

HOWELLS, J. and GREEN, A. (1988) Technological Innovation Structural Change and Location in UK Services, Gower, Aldershot.

HUDSON, R., RHIND, D. and MOUNSEY, H. (1984) An Atlas of EEC Affairs, Methuen, London.

HULL, C. (1983) 'Federal Republic of Germany', in Storey, D.J. (Ed.) The Small Firm : An International Survey, Croom Helm, London, 153-178.

HUTCHINSON, D. (1984) "LANS - technology and politics" Communications 1, 19-27.

HUWS, J. (1984) The New Homeworkers, Low Pay Unit, London.

INSTITUT FOR WIRTSCHAFTSFORSCHUNG (IFO) (1986) Service Industries : Role and Determinants of Producer Services in the Development of the European Economy" Preliminary Report to DGIII, Commission of the European Communities, Brussels.

ILLERIS, S. (1985) 'How to analyse the role of services in regional development', in Services - The New Economy : Implications for National and Regional Development. Proceedings of the Irish Branch of the Regional Studies Association 1985 Annual Conference, 26 March, Dublin.

ILLERIS, S. (1986) 'Classification of service activities', Working Document 3.3, 1986, FAST II programme, Commission of the European Communities.

INFORMATION DYNAMICS LTD. (1986) Impact of Interconnected National PTT Data Networks on the Users of European Information Services. Report for Directorate-General of Information Markets and Innovation, Luxembourg.

IRWIN, M.R. and MERENDA, M.J. (1987) "The network as corporate strategy" Transnational Data and Communications Report 10, 17-20.

JAPAN INFORMATION PROCESSING DEVELOPMENT CENTER (JIPDEC) (1987) Informatization White Paper - 1987 Edition JIPDEC, Tokyo.

JONES, P. (1981) "Retail innovations and diffusion" Area 13, 197-201.

JONES, P. (1985) "The growth of fast food operations in Britain". Geography 70, 347-350.

JONSCHER, C. (1983) "Information resources and economic productivity" Information Economics and Policy 1, 13-35.

KATOUZIAN, M.L. (1970) "The development of the service sectors : a new approach" Oxford Economic Papers 22, 362-382.

KEEBLE, D., OWENS, P. and THOMPSON, C. (1981) The Influence of Peripheral and Central Locations on the Relative Development of Regions, Department of Geography, University of Cambridge for Directorate-General for Regional Policy, Commission of the European Communities, Brussels.

KEEBLE, D. and WEAVER, E. (1986) 'Introduction' in Keeble, D. and Weaver, E. (Eds.) New Firms and Regional Development in Europe, Croom Helm, London, 1-34.

KEEBLE, D., OFFORD, J. and WALKER, S. (1986) Peripheral Regions in a Community of Twelve Member States : Final Report Department of Geography, University of Cambridge for Directorate-General for Regional Policy, Commission of the European Communities, Brussels.

KELLERMAN, A. (1984) "Telecommunications and the geography of metropolitan areas" Progress in Human Geography 8, 222-246.

KELLERMAN, A. (1985) "The evolution of service economies" The Professional Geographer 35, 133-143.

KORTE, W.B. (1986) 'Small and medium-sized establishments in Western Europe', in Keeble, D. and Wever, E. (Eds.) New Firms and Regional Development in Europe. Croom Helm, London, 35-51.

KOTTIS, G. (1985) 'The Service Sector in Greece' (mimeo) The Athens School of Economics and Business, Athens.

KROLLIS, H.P. (1986) 'Producer Services and Technological Change : The Internationalisation Issue' Research Centre for Urban and Regional Planning - TNO, Delft.

KROMMENACKER, R.J. (1987) 'La nature et les enjeux de la liberalisation multilaterale des services dans le contexte de la servisation des economies' Paper presented to Colloque Europrospective, La Villette, Paris, 23-25 April 1987.

LALOR, E. (1987) "Action for telecommunications development: STAR a European Community Programme" Telecommunications Policy 11, 115-120.

LAMBERTON, D.M. (1977) 'Structure and growth of the communications industry' in Tucker, K.A. (Ed.) The Economics of the Australian Service Sector. Croom Helm, London, 143-166.

LAMBERTON, D.M. (1983) 'Information economics and technological change' in MacDonald, S., Lamberton, D.M. and Mandeville, T.D. (Eds.) The Trouble with Technology : Explanations in the Process of Technological Change. Frances Pinter, London, 75-92.

LAMBOOY, J.G. and TORDOIR, P.P. (1985) 'Professional Services and Regional Development : A Conceptual Approach' in The Present and Future Role of Services in Regional Development : 16-18 October 1985, Seminar Proceedings - External Contributions Occasional Paper, No.74, FAST, Brussels.

LANGDALE, J. (1982) "Competition in telecommunications". Telecommunications Policy 6, 283-299.

LANGDALE, J. (1983) "Competition in the United States' long-distance telecommunications industry" Regional Studies 17, 393-409.

LANGDALE, J. (1983) "Electronic funds transfer and the internationalisation of the banking and finance industry" Geoforum 16, 1-13.

LANE, J.E. (1984) Electronic Funds Transfer : State of he Art Report National Computing Centre Ltd, Manchester.

LAZZARI, T. (1985) "Italian contribution - Electronic databases for small and medium-sized businesses" in Technical Change Centre 'Technical Tools for the New Services' Draft Final Report, SERV2 Programme : TCC, London (TCCR-85-008).

LAZZERI, Y., LEO, P.Y., MONNOYER, M.C. and PHILIPPE, J. (1985) 'Les Prestataires de Services, Les Petites et Moyennes Enterprises et Le Developpement Regional' in The Present and Future Role of Services in Regional Development : 16-18 October, Seminar Proceedings - External Contributions, Occasional Paper No.74, FAST, Brussels.

MACHLUP, F. (1962) The Production and Distribution of Knowledge in the United States. Princeton University Press, New Jersey.

MADDEN, D. (1986) "Irish eye service profit prospect" Computer Weekly 11 September, 1986.

MALECKI, E.J. (1977) "Firms and innovation diffusion : examples from banking" Environment and Planning A 9, 1291-1305.

MANDEVILLE, T. (1985) "Information flows between Australia and Japan", Papers of the Regional Science Association 56, 189-200.

MANDEVILLE, T., MacDONALD, S., THOMPSON, B. and LAMBERTON, D.M. (1983) Technology and Employment and the Queensland Information Economy, Report to the Department of Employment and Labour Relations, Queensland.

MARQUAND, J. (1979) 'The role of the tertiary sector in regional policy', Regional Policy Series 19, Collection Studies, Commission of the European Communities, Brussels.

MARSHALL, J.N. (1982a) "Linkages between manufacturing industry and business services" Environment and Planning A 14, 1523-1540.

MARSHALL, J.N. (1982b) 'Corporate Organisation of the Business Service Sector'. CURDS Discussion Paper No.43, University of Newcastle upon Tyne.

MARSHALL, J.N. (1985) "Business servies : the regions and regional policy" Regional Studies 19, 352-363.

MARSHALL, J.N. and BACHTLER, J.F. (1984) "Spatial perspectives on technological changes in the banking sector of the United Kingdom" Environment and Planning A 16, 437-450.

MARSHALL, J.N., DAMESICK, P. and WOOD, P. (1985) 'Understanding the location and role of producer services'. Paper presented at the Regional Science Conference, University of Manchester, September.

MASON, C.M. (1985) "The geography of 'successful' small firms in the United Kingdom" Environment and Planning A 17, 1499-1513.

MIDDLETON, R. (1985) "The development of the European information industry" Electronic Publishing Review 5, 49-60.

MILES, I. (1985) The Service Economy and Socio-economic Development, Science Policy Research Unit, University of Sussex. Draft paper for UNCTAD, December.

MINISTRY OF POSTS AND TELECOMMUNICATIONS (1987) Report on the Present State of Communications in Japan - Fiscal 1986 The Japan Times Ltd, Tokyo.

MOSS, M.L. (1986a) "Telecommunications and the future of cities" Land Development Studies 3, 33-44.

MOSS, M.L. (1986b) Telecommunications and International Financial Centres (mimeo), Graduate School of Public Administration, New York University, New York 10003.

MOSS, M.L. (1986c) Telecommunications Policy and World Urban Development. Paper presented to Annual Meeting of the International Institute of Communications, Edinburgh.

MOTHE, J.R. de la (1986) 'Financial Services' in Smith, A.D. (Ed.)
Technological Trends and Employment : 5. Commercial Service
Industries, Gower, Aldershot, 55-119.

NAIRN, G. (1987) "Data nets meet with discontent" Informatics 8, 46.

NEDERLANDS ECONOMISCH INSTITUT (1986) Telecommunications and the
Location of Producer Services in the Netherlands (TELOS). Final
Report for FAST II, Directorate General for Science, Research and
Development, Commission of the European Communities.

NICOL, L. (1985) 'Communications technology : economic and spatial
impacts' in Castells, M. (Ed.) High Technology, Space and Society
Sage Publications, Beverly Hills, 191-209.

NILLES, J.M., CARLSON, R., GREY, P. and HEINEMAN, G. (1976)
Telecommunications - Transportation Trade-Offs : Options for
Tomorrow, Wiley, New York.

NORA, S. and MINC, A. (1978) L'informatisation de la societe. La
Documentation Francaise, Paris.

NOYELLE, T.J. and STANBACK, T.M. (1984) The Economic Transformation of
American Cities, Rowman and Allanhold, Totawa, New Jersey.

NORTHERN IRELAND ECONOMIC COUNCIL (1982) Private Services in Economic
Development. Report No.30, NIEC.

NORTON, D. (1984) "Public policy for private sector services" Journal
of Irish Business and Administrative Research 6, 84-105.

OECD (1979) The Usage of International Data Networks in Europe.
Committee on Information, Computer Communication Policy (ICCP),
Series No.2, OECD, Paris.

OECD (1980) Policy Implications of Data Network Developments in the
OECD Area. Committee on Information, Computer Communications
Policy (ICCP), Series No.3, OECD, Paris.

OECD (1981) Information Activities, Electronics and Telecommunications
Technologies : Impact on Employment, Growth and Trade, Volume 1.
Committee on Information, Computer Communication Policy (ICCP),
Series No.6, OECD, Paris.

OECD (1985a) Update of Information Sector Statistics, Committee on
Information, Computer and Communications Policy (ICCP 84, 19),
OECD, Paris.

OECD (1985b) Software : An Emerging Industry Information, Computer and
Communications Policy Series No.9, OECD, Paris.

OECD (1986) Trends in the Information Economy. Committee on
Information, Computer and Communications Policy (ICCP), OECD,
Paris.

OFFICE OF TECHNOLOGY ASSESSMENT (1987) International Competition in Services OTA, Congress of the United States, US Government Printing Office, Washington DC.

PARSONS, G.F. (1972) "The giant manufacturing corporations and balanced regional growth in Britain", Area 4, 99-103.

PEAT, MARWICK, MITCHELL & CO. (1986) A Typology of Barriers to Trade in Services. Commission of the European Communities, Brussels.

PETIT, P. (19859 Automation of Services : The Case of the Banking Sector. DSTI/ICCP/84.19 OECD, Paris.

PHILLIPS, P. (1982) Regional Disparities, Lorimer, Toronto.

PIKE, N. and MOSCO, V. (1986) "Canadian consumers and telephone pricing" Telecommunications Policy 10, 17-32.

POLESE, M. (1981) "Interregional service flows, economic integration and regional policy; some considerations based on Canadian survey data" Revue d'Economic Regionale et Urbaine 4, 489-503.

POLESE, M. (1982) "Regional demand for business services and inter-regional service flows in a small Canadian region". Papers of the Regional Science Association 50, 151-163.

PORAT, M. (1977) The Information Economy : Definition and Measurement. United States Department of Commerce, Office of Telecommunications Special Publication, 77-12(i), Washington.

PRED, A.R. (1974) 'Industry, information and city-system inter-dependencies' in Hamilton, F.E.I. (Ed.) Spatial Perspectives on Industrial Organisation and Decision-Making, John Wiley, London, 105-39.

PRED, A.R. (1975) "Diffusion, organisational spatial structure and city system development", Economic Geography 51, 252-268.

PRED, A.R. (1977) City Systems in Advanced Economies, Hutchinson, London.

PRICE, D. (1987) "Costing a packet" The Guardian 12 March 1987.

PYE, R. and LAUDER, G. (1987) "Regional aid for telecommunications in Europe: a force for economic development" Telecommunications Policy 11, 99-113.

RADA, J.F. (1984) 'Development, telecommunications and the emerging service economy' Paper presented to the Second World Conference on Transborder Data Flow Policies, 26-29 June 1984, Rome.

RADNER, R. (1986) "The internal economy of large firms", Supplement to the Economic Journal 96. Conference Papers, 1-22.

RAJAN, A. (1984) New Technology and Employment in Insurance, Banking and Building Societies : Recent Experience and Future Impact Gower, Aldershot.

RAJAN, A. (1985) "Office technology and clerical skills", Futures 17, 410-413.

RAJAN, A. & PEARSON, R. (1986) UK Occupation and Employment Trends to 1990 : An Employer-Based Study of the Trends and their Underlying Causes, Butterworths, London.

READ, W.M. (1977) "Network control in global communications" Telecommunications Policy 1, 125-137.

REVESZ, R., DRESNER, S.H. and DIAZ, R. (1983) Transborder Data Flow : Issues, Barriers and Corporate Responses, Business International, New York.

ROBERTSON, J.A., BRIGGS, J.M. and GOODCHILD, A. (1982) Structure and Employment Prospects of the Service Industries. Department of Employment Research Paper 30, Department of Employment, London.

ROBINSON, P. (1983) "TDF : the hardy perennial" Telecommunications Policy 7 271-276.

ROBINSON, P. (1985) 'Transborder Data Flows : An Overview of Issues' in OECD (Ed.) Transborder Data Flows North Holland, Amsterdam, 15-29.

RODA (1980) The Impact of Micro-Electronics : A Tentative Appraisal of Information Technology. International Labour Office, Geneva.

RUBIN, M.R. and TAYLOR, E. (1981) "The US information sector and GNP : an input-output study "Information Processing and Management" 17, 163-194.

RUBIN, M.R. & HUBER, M.T. (1986) The Knowledge Industry in the United States, 1968-1980. Princeton University Press, Princeton, New Jersey.

RUGMAN, A.M. (1981) Inside the Multinationals : The Economics of Internal Markets. Croom Helm, London.

RUGMAN, A.M. (1982) 'Internationalisation and Non-Equity Forms of International Involvement' in Rugman, A.M. (Ed.) New Theories of the Multinational Enterprise, 9-23, Croom Helm, London.

SAPIR, A. and LUTZ, E. (1980) Trade in Non-Factor Services : Past Trends and Current Issues. WP-0410, World Bank Working Papers.

SAPIR, A. and LUTZ, E. (1981) Trade in Services : Economic Determinants and Development Issues. WP-0480, World Bank Working Papers.

SAUVANT, K.P. (1984) "Transborder data flows : importance, impact, policies" Information Services and Use 4, 3-30.

SAUVANT, K.P. (1986a) International Transactions in Services : The Politics of Transborder Data Flows Westview Press, Boulder Co.

SAUVANT, K.P. (1986b) Trade and Foreign Direct Investment in Data Services, Westview Press, Boulder Co.

SAUVANT, K.P. (1986c) "Trade in data services : the international context" Telecommunications Policy 10, 282-298.

SCHEMENT, J.R., LEIVROUW, L.A. and DORDICK, H.S. (1983) "The information society in California - social factors affecting its emergence" Telecommunications Policy 7, 64-72.

SCHUSTER, L. (1986) 'Role of innovation in the development of financial institutions' in Commission of the European Communities, Symposium on Europe and the Future of Financial Services: Proceedings/ Communications, DGXII(FAST) - DGXV, Brussels, 149-162.

SEMA (1984) Private Sector Perceptions of Community Interests in the Liberalisation of Trade in Services. Report to Directorate-General for External Relations, Commission of the European Communities, Brussels.

SEMA-METRA CONSEIL (1986a) Services to the Manufacturing Sector : A Long-Term Investigation : Report for FAST II, Directorate-General for Science, Research and Development, Commision of the European Communities, Brussels. FAST Occasional Paper 96.

SEMA-METRA CONSEIL (1986b) Services to the Manufacturing Sector : A Long-Term Investigation - Appendices. Report for FAST II, Directorate-General for Science, Research and Development, Commission for the European Communities, Brussels. FAST Occasional Paper 97.

SEMPLE, R.K. (1973) "Recent trends in the concentration of Corporate headquarters" Economic Geography 49, 309-318.

SEMPLE, R.K. (1977) "The spatial concentration of domestic and foreign multinational headquarters in Canada" Cahiers de Geographie de Quebec 21, 33-51.

SLEIGH, J., BOATWRIGHT, B., IRWIN, P. & STANYON, R. (1979) The Manpower Implications of Microelectronics Technology, HMSO, London.

SMITH, A.D. (1972) The Measurement and Interpretation of Service Output, NEDO, London.

SMITH, A. (1981) "The informal economy" Lloyds Bank Review 141, 45-61.

SMITH, A.D. (1986) 'Miscellaneous Services', in Smith, A.D. (Ed.) Technological Trends and Employment : 5. Commercial Service Industries, Gower, Aldershot, 1-53.

STANBACK, T.M. (1979) <u>Understanding the Service Economy : Employment, Productivity, Location</u>, John Haskins University Press, Baltimore.

STEFFENS, J. (1983) <u>The Electronic Office : Progress and Problems</u>. Policy Studies Institute Research Paper, 83-1.

STERN, E. and HOLTI, R. (1986) <u>Distance Working in Urban and Rural Settings</u>, FAST Distance Working Paper No.6, Tavistock Institute, London.

STIGLITZ, J.E. (1985) "Information and economic analysis : a perspective" <u>Supplement to the Economic Journal</u> 95, 21-41.

STONIER, T. (1986) "Towards a new theory of information" <u>Telecommunications Policy</u> 10, 278-281.

STOREY, D.J. and JOHNSON, S. (1986) <u>Employment and Occupational Structure in Smaller UK Businesses : Recent Trends and Projections to 1990</u> (mimeo) Centre for Urban and Regional Development Studies, University of Newcastle upon Tyne.

TAYLOR, M.J. (1975) "Organisational growth, spatial interaction and location decision-making" <u>Regional Studies</u> 9, 313-323.

TAYLOR, M.J. and HIRST, J. (1984) "Environment, technology and organisation: the restructuring of the Australian trading banks" <u>Environment and Planning</u> A16, 1055-1078.

TAYLOR, M.J. and THRIFT, N.J. (1981a) "British capital overseas : direct investment and corporate development in Australia" <u>Regional Studies</u> 15, 183-212.

TAYLOR, M.J. and THRIFT, N.J. (1981b) Organisation, location and political economy : towards a geography of business organisations. <u>CURDS Discussion Paper</u> No.38, University of Newcastle upon Tyne.

TAYLOR, M.J. and THRIFT, N.J. (1983) "Business organisations, segmentation and location" <u>Regional Studies</u> 17, 445-465.

TAYLOR, M.J. and THRIFT, N.J. (1986) 'Introduction : new theories of multinational corporations' in Taylor, M. and Thrift, N.J. (Eds.) <u>Multinationals and the Restructuring of the World Economy</u>, 1-20, Croom Helm, London.

THOMPSON, G. (1971) "Greening of the wired City" <u>Telesis</u>, Bell Northern Research, Ottawa.

THORNGREN, B. (1970) "How do contact systems affect regional development?" <u>Environment and Planning</u> A 2, 409-427.

THRIFT, N.J. (1985) "Research policy and review. Taking the rest of the world seriously? The state of British urban and regional research in a time of economic crisis" <u>Environment and Planning</u> A 17, 7-24.

TODD, P.C. and STRASSER, R.R. (1985) Electronic Financial Information Services in Europe. Report by Information Dynamics Ltd. to the Directorate General for Information Market and Innovation. Commission of the European Communities, Brussels.

UNCTAD (1983) International Services Transactions. United Nations Conference on Trade and Development (UNCTAD) Secretariat, April.

UNITED NATIONS (1983) Transborder Data Flows : Access to the International On-Line Data Base Market. Centre on Transnational Corporations, New York.

van HASELEN, H., MOLLE, W. and de WIT, R. (1985) Technological Change and Service Employment in the Regions of Europe : The Case of the Banking and Insurance Sector. Paper presented at 'Technological Change and Employment : Urban and Regional Dimensions' 28-30 March 1985, Zandvoort.

WESTAWAY, J. (1974a) "Contact potential and the occupational structure of the British urban system 1961-1966 : an empirical study" Regional Studies 8, 57-73.

WESTAWAY, J. (1974b) "The spatial hierarchy of business organisations and its implications for the British Urban System" Regional Studies 8, 145-155.

WEVER, E. (1986) 'New firm formation in the Netherlands' in Keeble, D. and Wever, E. (eds.) New Firms and Regional Development in Europe, Croom Helm, London, 54-74.

WILLIAMS, M.E. (1984) "Policy issues for electronic databases and database systems " The Information Society 2, 3-4.

WILLMAN, P. and COWAN, R. (1984) 'New Technology in Banking : The Impact of Antotellers on Staff Numbers' in Warner, M. (Ed.) Microprocessors, Manpower and Society, Gower, Aldershot.

WILLIAMSON, O.E. (1975) Markets and Hierachies : Analysis and Antitrust Implications. The Free Press, New York.

WOOD, P.A. (1984) "Regional industrial development" Area 16, 281-289.

WOOD, P.A. (1986) "The anatomy of job loss and job creation : some speculations on the role of the 'producer service' sector" Regional Studies 20, 37-46.

Appendix 1

1.   Introduction

This appendix provides an assessment and analysis of the regional distribution of service employment between 1977 and 1983, the latest period for which data are available. Unfortunately for the more detailed NACE service sector analysis the latest time period for which data were available was 1977-1981. The areal base unit for this study is level II or equivalent region (see Commission of the European Communities 1984, 2-3). The data for Europe 10 is based on the Labour Force Survey except for Greek 1977 figures. For Spain data were taken from the national "Encuesta de Poblacion Activa" from INE, which is comparable with Eurostat data, whilst data for Portugal was taken from Eurostat "Employment and Unemployment Yearbook" of 1985. Gaps, however, still remain in the data. No data on a temporal basis (Section 4) were available for Spain, Portugal, or Greece for service activities disaggregated on the NACE classification system (NACE 6-9; see Appendix 2). Moreover care should be made in interpreting service employment change as an indicator of growth in overall service activity. Thus a lack of growth (or decline) in service jobs in a region may reflect a high level of labour productivity growth arising from capital investment and technological change rather than static (or declining) conditions in overall service activity (e.g. output and investment) in such a locality.

Nonetheless, despite these weaknesses the data offers a valuable picture of service employment in the European Community, representing the only source of information relating to services available on a disaggregated areal basis, suitable for spatial analysis.

The rest of this appendix will be organised as follows : Section 2 will provide a brief overview in trends in total service employment ; Section 3 presents a broad examination of service structure change at level II regions; whilst Section 4 will conclude by investigating locational trends of the main NACE service categories. As such Sections 2 and 3 extends some of the analysis of regional service employment change between 1973-79 and 1979-83 undertaken by Keeble et al. (1981; 1986) as part of their wider study on centrality and peripherality within the European Community; (see also Hudson, Rhind and Mounsey 1984, 67-72 and 88-99).

2.   Regional Trends in European Service Employment, 1977-1983

Figures 1 and 2 record trends in the regional distribution of service employment in the European Community between 1977-1983 on an absolute and percentage basis. What is evident from these maps is that in both absolute, and more particularly relative terms, some of the highest rates of service employment growth are in the peripheral, less favoured regions of the Community. Thus Ireland, Portugal, southern Italy and more peripheral and rural parts of France exhibit some of the strongest areas of service employment expansion. This confirms the core-periphery, regional trend observed by Keeble et.al. (1981, 133; 1986, 93) during the 1970s and early 1980s where service employment in

FIGURE 1

EUROPEAN COMMUNITY ABSOLUTE CHANGE IN EMPLOYEES IN SERVICES BETWEEN 1977 AND 1983

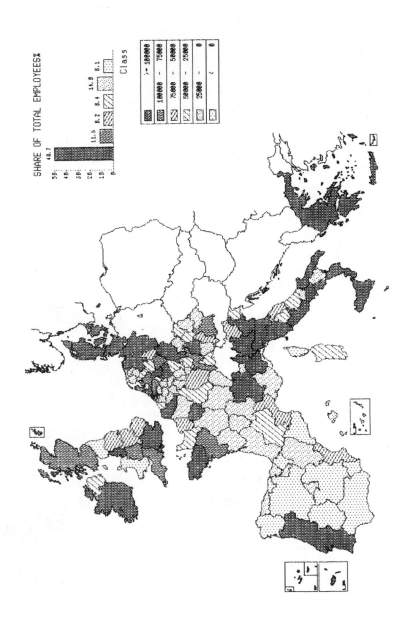

FIGURE 2

EUROPEAN COMMUNITY CHANGE IN SERVICE EMPLOYEES BETWEEN 1977 AND 1983 AS A PERCENTAGE OF 1977 LEVELS

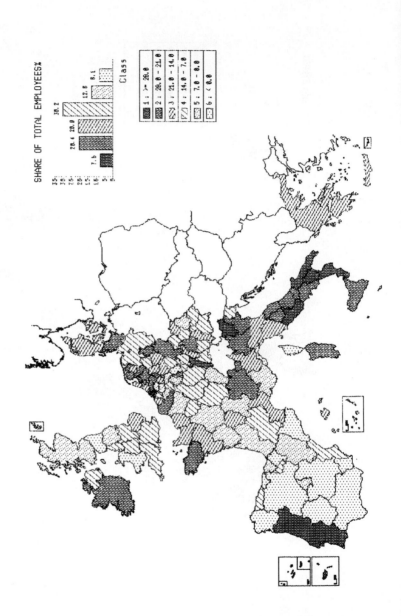

FIGURE 3

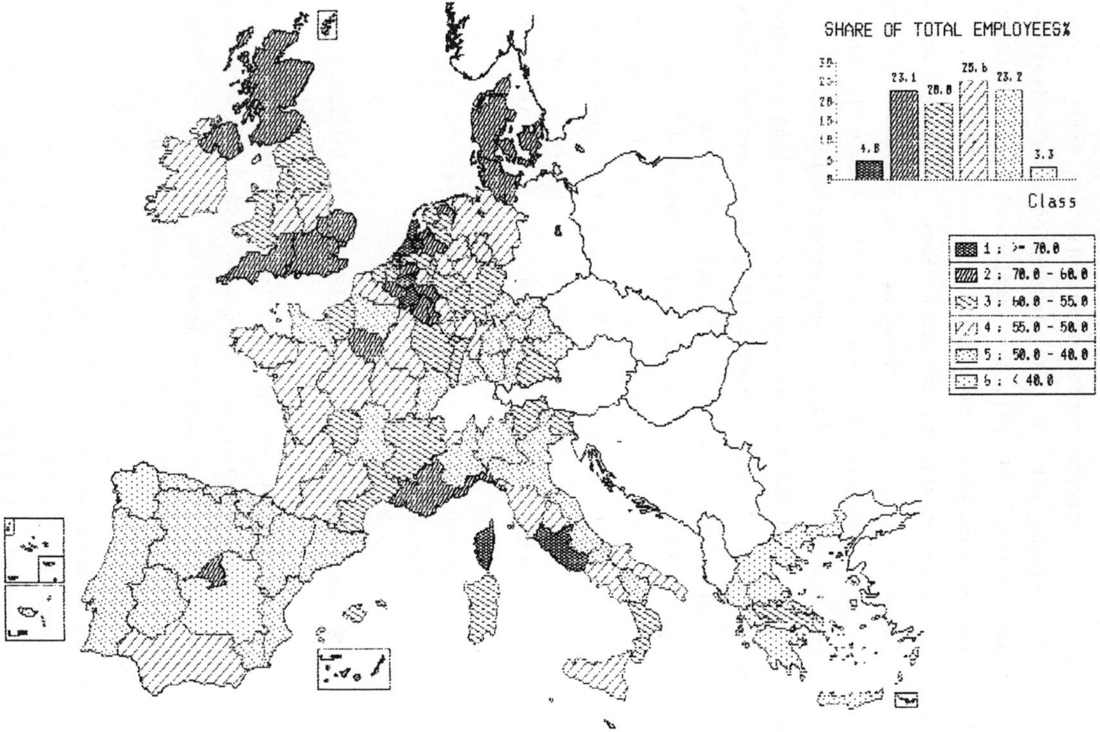

EUROPEAN COMMUNITY SERVICE EMPLOYEES AS A PERCENTAGE OF TOTAL EMPLOYEES 1983

peripheral regions grew more rapidly and by a greater volume of jobs than either the central or intermediate regions.

In relation to service sector specialisation there appear to be two types of regions with a high degree of specialisation in service employment in 1983. Firstly, core regions which dominate their national urban and regional hierarchies (such as Madrid in Spain, the South East (London) region of the UK, or Lazio (Rome) in Italy) and secondly, by contrast, more peripheral regions such as Northern Ireland, Scotland, Wales, Corsica and Provence and Denmark (Figure 3; see also Hudson, Rhind and Mounsey 1984, 93). These latter economies are characterised by a high level of public, consumer sector services in comparison to the former areas which are dominated by private sector producer services (Section 3; Figure 5). On the basis of those regions exhibiting increasing specialisation in services between 1977 and 1983 no clear regional trends are apparent (Figure 4) apart from the fact that the regions exhibiting the strongest shift towards a service orientation are predominantly intermediate regions and that major metropolitan core regions are localities which are relatively static in service employment change rates. This latter phenomenon may be due to above average rates of labour productivity growth in the types of service activities concentrated there, i.e. producer services.

3.   Regional Patterns of Service Structure in Europe, 1983

Regional analysis of service structure is based upon a simple index devised by Keeble, Owen and Thompson (1981) which measures regional employment in a given year in producer services to that in consumer services. Obviously this bi-polar classification of services is crude and inexact, with for example producer services (= NACE 7 + NACE 8) including a consumer component, whilst consumer services (= NACE 6 + NACE 9) covering a number of producer service components. However, this simple categorisation can be a helpful indicator of the wider, longer term growth potential in services of regions. It might be suggested that growth in producer services is of greater economic value to a region, in the sense of the potential for generating exogenous income

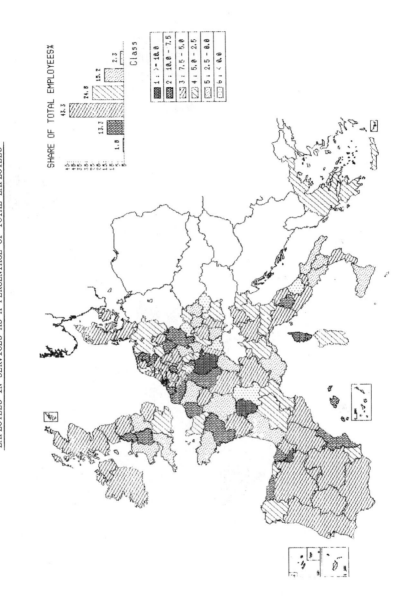

SHARE OF TOTAL EMPLOYEES%

| 1.0 | 13.2 | 43.3 | 24.8 | 15.7 | 2.3 |

Class

1 : >= 10.0
2 : 10.0 - 7.5
3 : 7.5 - 5.0
4 : 5.0 - 2.5
5 : 2.5 - 0.0
6 : < 0.0

FIGURE 5

SERVICES STRUCTURE INDEX 1983

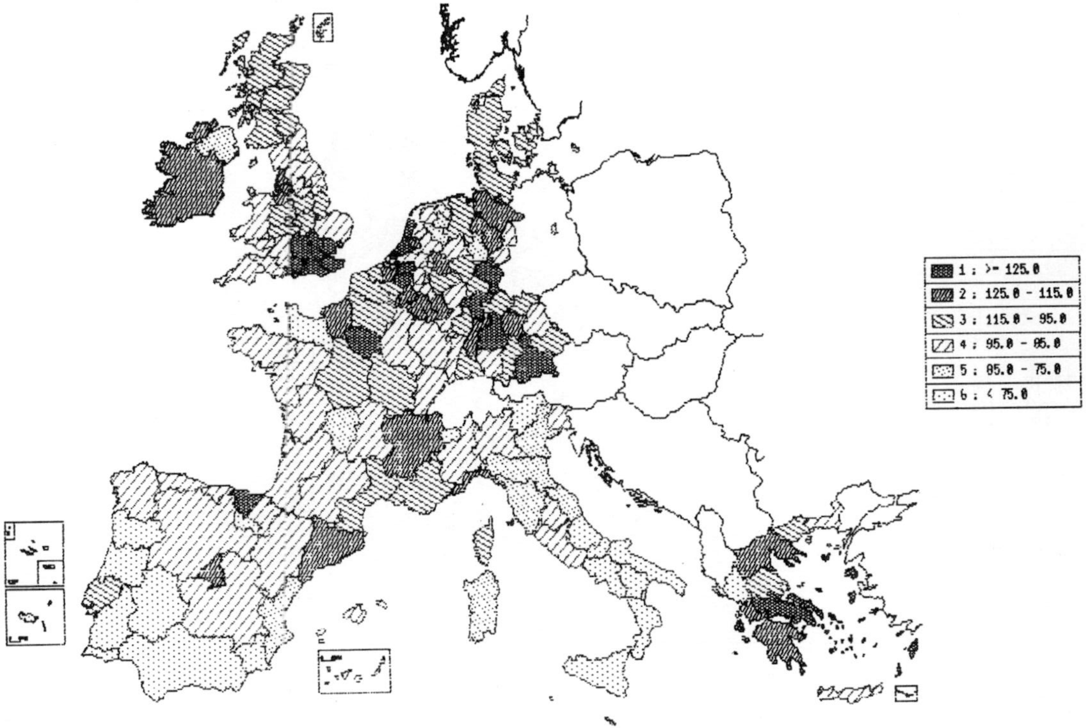

1 : >= 125.0
2 : 125.0 - 115.0
3 : 115.0 - 95.0
4 : 95.0 - 85.0
5 : 85.0 - 75.0
6 : < 75.0

and strengthening the competitive efficiency of a region, than a similar increment in consumer service employment. Thus a relatively high producer–consumer ratio may be an indication of a more favourable long-term service industry structure.

In relation to the regional pattern of service structure in 1983 (Figure 5) it is apparent that the core regions, centred on Paris, Athens and London, together with a scatter of other dynamic, core regions in Belgium, the Netherlands and Germany had the highest producer–consumer ratios. Other intermediate centres, for example, in Pais Vasco (Spain), the North West (UK), Rhone-Alps (France) and Ireland, however, also had relatively high producer–consumer ratios. Overall therefore it was generally the more dynamic, growth oriented regions that had the greatest orientation towards producer services and this encompassed both core and intermediate regions.

In the context of shifts in service structure over the period, 1977–1981 only limited data were available. However, there was some suggestion that the spatial change pattern in service structure was one of both national and regional components. Evidence of a national component was most strongly evident in the UK where the service structure for all of its regions were moving against the Community average, with a noticeable shift towards more locally oriented consumer services. This is likely to have been maintained during the 1980s in the UK with the continuation of consumer–led growth. By contrast France and Germany were undergoing increasing specialisation in more export oriented, producer services. However, on a regional level, in those countries which were becoming more specialised in producer services, particularly France and Italy, it was generally the more peripheral, less–urbanised areas which were experiencing the greatest relative producer–consumer service shifts.

4. Regional Trends in Service Sector Specialisation within the European Community, 1977–1981

This section will outline in brief regional trends in the degree of specialisation in the following service sector components (Appendix 2) :

4.1 NACE 6 Distribution Trades, Hotels, Catering and Repairs

4.2 NACE 7 Transport and Communication

4.3 NACE 8 Banking & Finance, Insurance, Business Services and Renting

4.4 NACE 9 Other Services

4.1 Regional Specialisation Trends in Distribution and Hotel Services

In relation to specialisation in distribution and hotel services it was
the key tourist areas of Greece and northern Italy which had a high
percentage of the total service employment working in these sectors
(Figure 6A).   In terms of relative (percentage change) growth in
distribution and hotel services there were no clear regional trends,
although in France and Italy there was a number of the above average
regions were in more peripheral and rural localities (Figure 6B).

Figure 6C indicates those regions which are becoming more specialised
within distribution and hotel services in relation to service employment
overall.   However, no clear regional pattern emerges in terms of shifts
in specialisation, except those regions which were becoming more
specialised in hotels and distribution tended to be more intermediate or
peripheral areas.

FIGURE 6a

EUROPEAN COMMUNITY EMPLOYMENT IN DISTRIBUTION AND HOTELS AS A PERCENTAGE OF TOTAL SERVICE EMPLOYMENT 1981

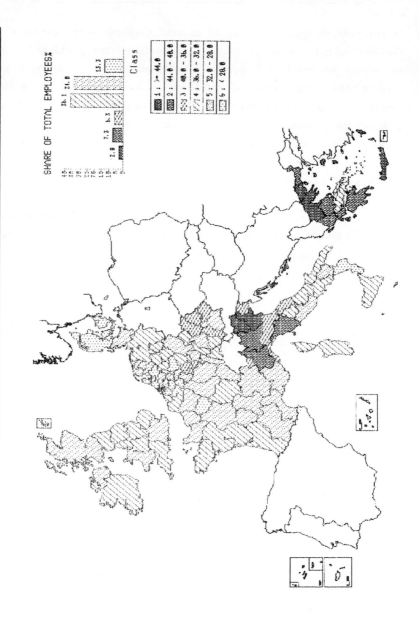

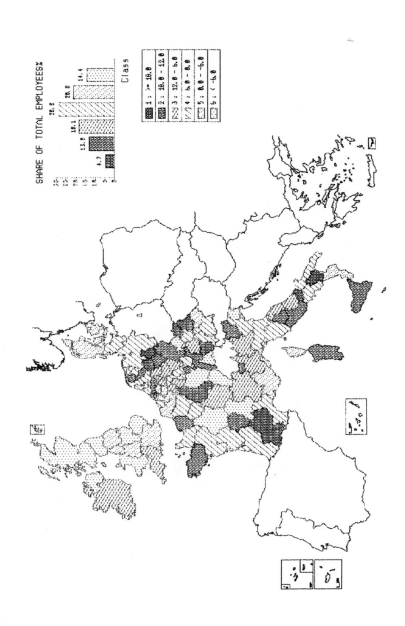

FIGURE 6c

EUROPEAN COMMUNITY PERCENTAGE POINT DIFFERENCE 1977-81 IN DISTRIBUTION AND HOTEL EMPLOYMENT AS A

PERCENTAGE OF TOTAL SERVICE EMPLOYMENT

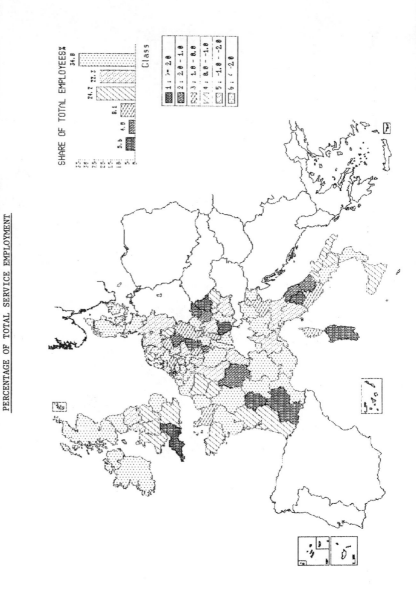

4.2 Regional Specialisation Trends in Transport and Communication Services

In the context of transport and communication no clear regional patterns are evident in terms of over- and under-concentration in this service group (Figure 7A) apart from the national phenomenon of Greece with its marked service orientation in this sector. In relation to employment change between 1977 and 1981 the highest increases were found in France and in a number of regions bordering onto France (Figure 7B). Again in terms of increasing employment specialisation in this sector it was the French regions which dominated, with other regions recording an increasing orientation in the transport and communication located in Belgium and parts of Germany (Figure 7C).

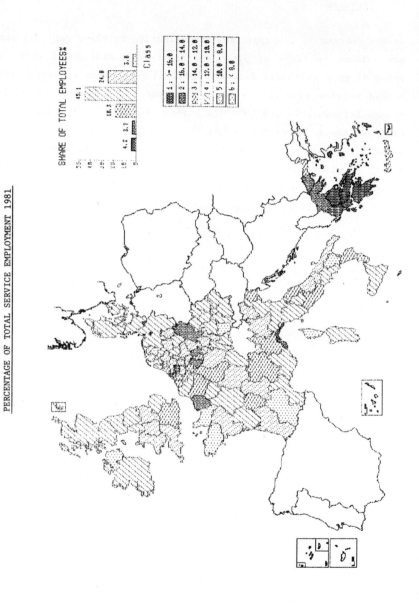

FIGURE 7a

EUROPEAN COMMUNITY EMPLOYMENT IN TRANSPORT AND COMMUNICATIONS AS A

PERCENTAGE OF TOTAL SERVICE EMPLOYMENT 1981

214

FIGURE 7b

EUROPEAN COMMUNITY CHANGE IN EMPLOYMENT IN TRANSPORT AND COMMUNICATIONS 1977-1981 AS A

PERCENTAGE OF 1977 LEVELS

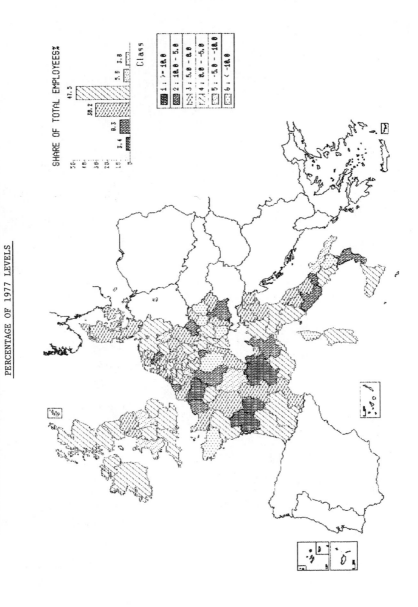

FIGURE 7c

EUROPEAN COMMUNITY PERCENTAGE POINT DIFFERENCE 1977-81 IN TRANSPORT AND

COMMUNICATIONS EMPLOYMENT AS A PERCENTAGE OF TOTAL SERVICE EMPLOYMENT

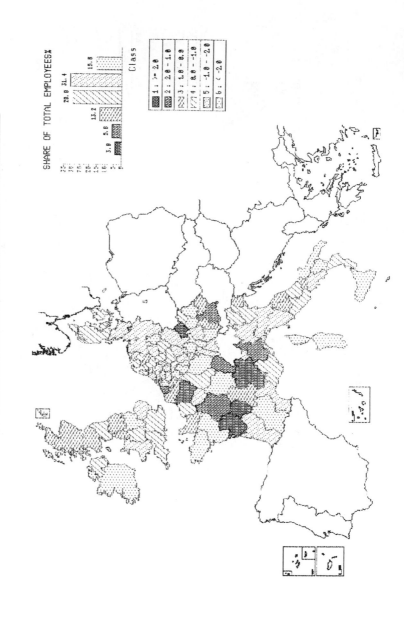

4.3 Regional Specialisation Trends in Financial and Business Services

Not unexpectedly key financial and business centres in and around
London, Paris, Amsterdam and Brussels were all evident in terms of
over-representation in these service activities (Figure 8A). Above
average levels in these services are also evident in southern and
central Germany, south-eastern France and Ireland. However, in terms of
employment growth rates in financial and business services it was in
regions generally near to or adjacent the major financial centres which
appeared to benefit most from above average growth rates. Level II
regions which had key financial centres located within them experienced
static or declining employment change, perhaps reflecting a more rapid
take-up of new technology and consequent increases in labour
productivity (Figure 8B). Those regions exhibiting increasing
specialisation in financial business and related service activities were
Ireland, and large parts of the UK and Germany, together with northern
Italy and scattered areas in France (Figure 8C). Some of the increased
specialisation in parts of the more peripheral regions of the Community
may reflect a simple catching up process in terms of the local provision
of financial and business services.

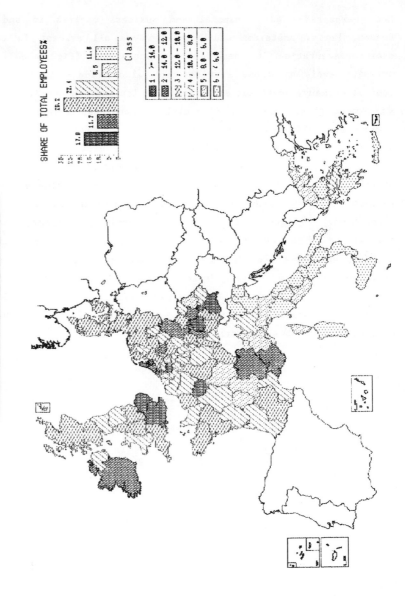

FIGURE 8a

EUROPEAN COMMUNITY EMPLOYMENT IN FINANCE AND BUSINESS SERVICES AS A

PERCENTAGE OF TOTAL SERVICE EMPLOYMENT 1981

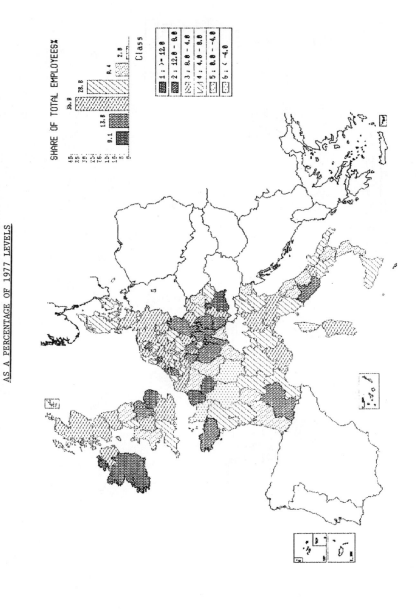

FIGURE 8b

EUROPEAN COMMUNITY CHANGE IN EMPLOYMENT IN FINANCE AND BUSINESS SERVICES 1977 – 1981

AS A PERCENTAGE OF 1977 LEVELS

219

FIGURE 8c

EUROPEAN COMMUNITY PERCENTAGE POINT DIFFERENCE 1977-81 IN FINANCE AND

BUSINESS SERVICES EMPLOYMENT AS A % OF TOTAL SERVICE EMPLOYMENT

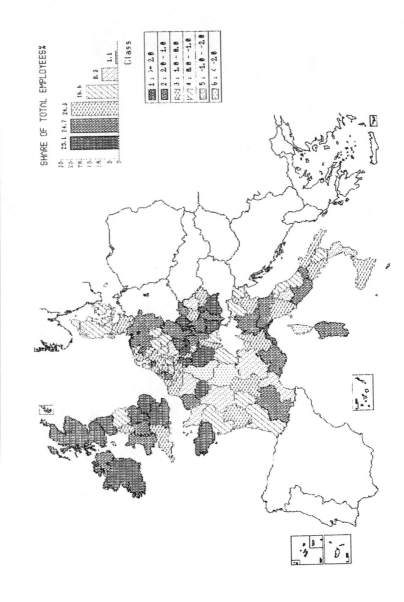

220

4.4 Regional Specialisation Trends in Other Services

Above average concentration ratios in other services, which covers a range of mainly public and professional consumer services (Appendix 2), are centred in localities in southern Italy, Northern Ireland, northern Denmark, southern Belgium and less urbanised areas of France (Figure 9A). In terms of employment change in these services, regions with high growth rates are centred in southern most parts of Italy, and a scatter of regions in France, Belgium and the Netherlands (Figure 9B). Strong national components (and policy orientation) are revealed in the above average growth rates in Denmark and Ireland, whilst, by contrast, the UK has experienced a net decline in these primarily public-related service activities. These regional changes are also evident in the shifts in orientation towards basic sectors, although a number of regions in France also record an increasing public and professional consumer service orientation perhaps associated with an under-performance in growth in other service sectors.

FIGURE 9a

EUROPEAN COMMUNITY EMPLOYMENT IN OTHER SERVICES AS A PERCENTAGE OF TOTAL SERVICE EMPLOYMENT 1981

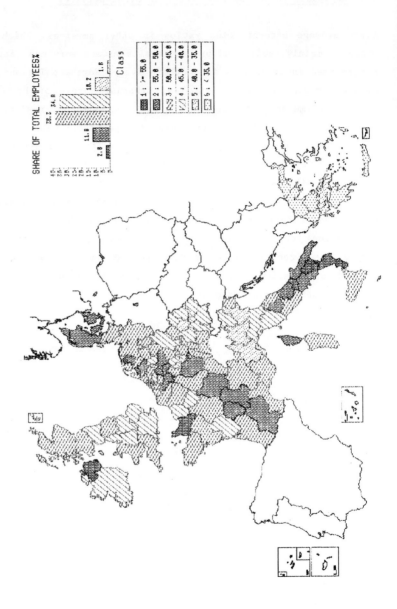

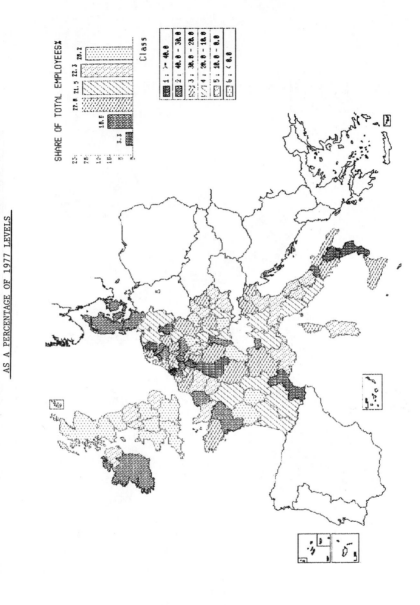

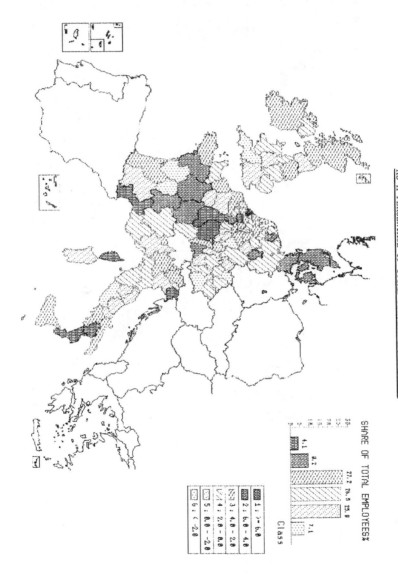

FIGURE 9c

EUROPEAN COMMUNITY PERCENTAGE POINT DIFFERENCE BETWEEN 1977–1981 IN OTHER SERVICES EMPLOYMENT

AS A PERCENTAGE OF TOTAL SERVICE EMPLOYMENT

Appendix 2

SUMMARY OF DIVISIONS AND CLASSES OF N.A.C.E.

0. AGRICULTURE, HUNTING, FORESTRY AND FISHING

    01   Agriculture and hunting
    02   Forestry
    03   Fishing

1. ENERGY AND WATER

    11   Extraction and briquetting of solid fuels
    12   Coke ovens
    13   Extraction of petroleum and natural gas
    14   Mineral oil refining
    15   Nuclear fuels industry
    16   Production and distribution of electricity, gas, steam and hot water
    17   Water supply: collection, purification and distribution of water

2. EXTRACTION AND PROCESSING OF NON-ENERGY PRODUCING MINERALS AND DERIVED PRODUCTS; CHEMICAL INDUSTRY

    21   Extraction and preparation of metalliferous ores
    22   Production and preliminary processing of metals
    23   Extraction of minerals other than metalliferous and energy-producing minerals; peat extraction
    24   Manufacture of non-metallic mineral products
    25   Chemical industry
    26   Man-made fibres industry

3. METAL MANUFACTURE; MECHANICAL, ELECTRICAL AND INSTRUMENT ENGINEERING

    31   Manufacture of metal articles (except for mechanical, electrical and instrument engineering and vehicles)
    32   Mechanical engineering
    33   Manufacture of office machinery and data processing machinery
    34   Electrical engineering
    35   Manufacture of motor vehicles and of motor vehicle parts and accessories
    36   Manufacture of other means of transport
    37   Instrument engineering

4. OTHER MANUFACTURING INDUSTRIES

    41/42   Food, drink and tobacco industry
    43   Textile industry
    44   Leather and leather goods industry (except footwear and clothing)
    45   Footwear and clothing industry
    46   Timber and wooden furniture industries
    47   Manufacture of paper and paper products; printing and publishing
    48   Processing of rubber and plastics
    49   Other manufacturing industries

5. BUILDING AND CIVIL ENGINEERING

    50   Building and civil engineering

6. DISTRIBUTIVE TRADES, HOTELS, CATERING, REPAIRS

    61   Wholesale distribution (except dealing in scrap and waste materials)
    62   Dealing in scrap and waste materials
    63   Agents
    64/65   Retail distribution
    66   Hotels and catering
    67   Repair of consumer goods and vehicles

7. TRANSPORT AND COMMUNICATION

    71   Railways
    72   Other land transport (urban transport, road transport etc)
    73   Inland water transport
    74   Sea transport and coasting shipping
    75   Air transport
    76   Supporting services to transport
    77   Travel agents, freight brokers and other agents facilitating the transport of passengers or goods; storage and warehousing
    79   Communication

8. BANKING AND FINANCE, INSURANCE, BUSINESS SERVICES, RENTING

    81   Banking and finance
    82   Insurance except for compulsory social insurance
    83   Activities auxiliary to banking and finance and insurance; real estate transactions (except letting of real estate by the owner), business services
    84   Renting, leasing and hiring of movables
    85   Letting of real estate by the owner

9. OTHER SERVICES

    91   Public administration, National defence and compulsory social security
    92   Sanitary services and administration of cemeteries
    93   Education
    94   Research and development
    95   Medical and other health services; veterinary services
    96   Other services provided to the general public
    97   Recreational services and other cultural services
    98   Personal services
    99   Domestic services
    00   Diplomatic representation, international organisations and allied armed forces

Source: "General Industrial Classification of Economic Activities within the European Communities", Eurostat, 1970.